WELCOME

Welcome to this visual celebration of a decade that, for many of us, represents the high-water mark of naval power. As you open these pages, you are stepping into a world of steam, salt spray, and the thunder of afterburners—a world that has captivated me for as long as I can remember.

There is an inherent drama to a carrier deck; it is a choreographed chaos that demands perfection, and the aircraft that inhabited that space in the 1980s were some of the most iconic machines ever built.

One of the primary reasons I felt compelled to produce this work is how perfectly the subject reflects the geopolitical situation of the era. The 1980s were the final, soaring crescendo of the Cold War. The deployment of the carrier air wing to areas of conflict and tension makes this subject a historical testimony to the US involvement overseas. This tension gave the aircraft a sense of purpose and "weight" that makes the subject truly fascinating to study.

As an aviation artist, this decade holds a special, almost romantic significance. We witnessed a dramatic aesthetic shift: the transition from the colourful, high-visibility paint schemes of the 1960s and 1970s—with their vibrant squadron insignia and white bellies—to the "grey ghosts" clad in the Tactical Paint Scheme (TPS). Capturing the way light plays off those low-visibility, weathered greys is a unique challenge.

I want to be clear that this book is not intended to be an in-depth historical document or a dense technical manual. Instead, it is a curated overview of the significant events and the legendary airframes that defined the decade. My goal is to capture the "feel" of the 80s fleet, providing a gateway for enthusiasts and a nostalgic journey for veterans. It is my sincere hope that this volume will be the first of many, followed by similar books exploring the unique character of other decades.

I am immensely proud to put my signature on this publication. It represents a lifetime of passion, and I hope it ignites that same spark in you.

Mads Bangsø

An E-2C Hawkeye of VAW-123 "Screwtops" over the Mediterranean Sea in 1986.

CONTENTS

ISBN: 978 1 83632 230 6
Editor: Mads Bangsø
Senior editor, specials: Roger Mortimer
Email: roger.mortimer@keypublishing.com
Cover Design: Panda Media
Design: Panda Media
Advertising Sales Manager: Sam Clark
Email: sam.clark@keypublishing.com
Tel: 01780 755131
Advertising Production: Becky Antoniades
Email: Rebecca.antoniades@keypublishing.com

SUBSCRIPTION/MAIL ORDER
Key Publishing Ltd, PO Box 300, Stamford, Lincs, PE9 1NA
Tel: 01780 480404
Subscriptions email: subs@keypublishing.com

Mail Order email: orders@keypublishing.com
Website: www.keypublishing.com/shop

PUBLISHING
Group CEO: Adrian Cox
Publisher: Steve O'Hara

Published by
Key Publishing Ltd, PO Box 100, Stamford, Lincs, PE9 1XQ
Tel: 01780 755131 **Website:** www.keypublishing.com

PRINTING
Precision Colour Printing Ltd, Haldane, Halesfield 1, Telford, Shropshire. TF7 4QQ

DISTRIBUTION
Frontline Distribution Solutions Ltd, 2 Poultry Avenue, London, EC1A 9PU
Enquiries: 02074 294000 or info@flgroup.co.uk

IN THE BE

For the United States Navy, the 1980s was a decade defined by radical transformation. Emerging from the 1970s—an era focused on the hard-won lessons of Vietnam—the Navy found itself thrust into a global landscape of escalating tension. The Cold War had evolved into a high-stakes standoff with the Soviet Union, opening new maritime fronts from the North Atlantic to the Pacific. Simultaneously, rising volatility across the Middle East and North Africa demanded the constant, watchful presence of Carrier Air Wings.

This was the decade when the "600-ship Navy" became a reality, and the nuclear-powered *Nimitz*-class supercarriers began to replace the aging conventional giants of the 1950s. But the evolution wasn't just in the hulls; it was in the very soul of naval aviation.

The 1980s occupied a unique moment in time—a "sweet spot" where the analoguepast met the digital future. On a single deck, one could witness the veteran F-4 Phantom II, a "heavy-metal" legend of the 1960s, launching alongside the F/A-18 Hornet, a high-tech marvel that ushered in the era of the multi-role strike fighter.

It was the era of the Grumman F-14 Tomcat in its absolute prime—the "Top Gun" years—where swing-wing fleet defenders stood as the first line of defence against a Soviet bomber threat that felt both real and imminent. It was also the twilight of the "A-jets"—the A-6 Intruder and A-7 Corsair II. These rugged workhorses provided the heavy-hitting punch of the air wing just before the dawn of precision-guided dominance forever changed the face of strike warfare.

This book is more than a catalogue of airframes; it is a chronicle of a decade where the Navy served as the primary instrument of American foreign policy. From the "Line of Death" in the Gulf of Sidra to high-stakes intercepts over the North Atlantic, the 1980s Carrier Air Wing was constantly in the thick of the action. We explore the significant events that defined the era, showing how these machines and their crews were tested in the real-world flashpoints that dominated the evening news.

To truly understand 1980s naval aviation, one must learn its language. Why were some aircraft adorned in vibrant "Easter egg" liveries while others wore ghost-like Tactical Paint Schemes? What did the "double-nuts" on a nose signify, and why did a pilot's rank dictate their assigned modex number?

In the following pages, we go beneath the surface. We break down the "anatomy" of the air wing, explaining the logic behind the numbering, the tradition of the tail codes, and the evolution of the markings that made every squadron a unique tribe within the fleet.

The 1980s was the cauldron in which modern naval power was forged. Whether you are a modeler searching for the perfect shade of "Gull Gray," a historian seeking an overview of significant events, or an enthusiast nostalgic for the roar of screaming turbofans and the snap of steam catapults, this book is your boarding pass.

Welcome back to the deck of a 1980s supercarrier. You're on the on-ramp to the Highway to the Danger Zone!

ABOVE: F-14A Tomcat (160661) from VF-51 "Screaming Eagles" from the August 12, 1986 to February 5, 1987 deployment aboard USS Carl Vinson (CVN-70) to the Westpac. and Indian Ocean. The 1980s were, in many ways, the decade of the F-14 Tomcat, as the aircraft slowly but steadily replaced the aging F-4 Phantom II and became the backbone of the U.S. Navy's fighter force. For the general public, the Tomcat also became synonymous with naval aviation following the release of Top Gun in 1986, in which aircraft from VF-51 and VF-111 were repainted and featured prominently as the true stars of the film

GINNING

Aircraft from CVW-9 fly over USS Kitty Hawk (CV-63) during the January 3, 1987 to June 29, 1987 around the world deployment.
(PHOTO: OFFICIAL US NAVY PHOTO BY PH3 BOTTORF)

A-7E Corsair (158830) from VA-37 "Ragin´ Bulls" from the March 10, 1980 to August 27, 1980 deployment aboard USS Saratoga (CV-60) in the Mediterranean Sea. The SLUF, a veteran of the Vietnam War, would soldier on for another decade as the 1980s began, serving as the backbone of the U.S. carrier-borne attack force. The aircraft would end its service life in a landing mishap aboard USS John F. Kennedy (CV-67) on January 24, 1991.

THE CARRIER AIR WING ENTERS THE 1980s

During the 1980s, the typical structure of a Carrier Air Wing consisted of two fighter squadrons, designated as the first and second squadrons within the wing. These fighter squadrons typically carried modex numbers in the 1xx and 2xx ranges, respectively. They were equipped with either the F-14 Tomcat or the F-4 Phantom II, the latter of which was phased out during the first half of the decade.

The third, fourth, and fifth squadrons were typically attack squadrons. Units operating the A-7 Corsair II were assigned modex numbers in the 3xx and 4xx ranges, while A-6 Intruder-equipped squadrons fell within the 5xx range. The 6xx range was generally shared among the Airborne Early Warning Squadron (VAW), the Tactical Electronic Warfare Squadron (VAQ), and the Helicopter Anti-Submarine Squadron (HS). Within this range, the E-2 Hawkeye-equipped VAW squadron usually used the first four numbers (600–603), followed by the EA-6B Prowler-equipped VAQ squadron (604–607), and the HS squadron starting from 610 onward.

A typical air wing also included an Anti-Submarine Squadron (VS), which operated the S-3 Viking and typically carried modex numbers in the 7xx range. In some cases, a Fleet Air Reconnaissance Squadron (VQ) was also part of the air wing, operating the EA-3B Skywarrior during this period.

Although this structure was common, notable deviations occurred throughout the 1980s. One significant deviation from the above, described structure was the introduction of the F/A-18 Hornet and the formation of Strike Fighter Squadrons (VFA). In some instances, the VFA units replaced the two traditional VA squadrons; in others, three VFA squadrons were substituted for both the fighter squadrons and the Corsair II-equipped attack squadrons. This development represented a major shift in air wing composition, which had previously maintained distinct fighter and attack squadrons since the Second World War.

USS *Carl Vinson* (CVN-70) flight deck during the August 12, 1986 to February 5, 1987 deployment to the Westpac. and Indian Ocean. In this photo is seen A-7E Corsair IIs from VA-27 "Royal Maces", A-6E Intruders of VA-52 "Knightriders", EA-6B Prowlers of VAQ-134 "Garudas" and a single C-2A Greyhound COD.
(PHOTO: OFFICIAL US NAVY PHOTO BY PH2 EDDIE E. CORDERO)

F-14A Tomcat (160687) of VF-51 "Screaming Eagles" from the May 20, 1979 to February 25, 1980 deployment aboard USS Kitty Hawk (CV-63) in the westpac. The Tomcat was quickly becoming the mainstay of the U.S. carrier-borne fighter force even before the 1980s began. It was a far more advanced and capable platform than its predecessor, in part due to the AIM-54 Phoenix missile that this aircraft is carrying.

F-14A Tomcat (162701) of VF-14 "Tophatters" from the August 2, 1988 to February 1, 1989 deployment aboard USS John F. Kennedy (CV-67) to the Mediterranean Sea. Fighter Squadron 14 had transitioned to the Grumman F-14A Tomcat in the mid-1970s and, by 1988, had become one of the most experienced Tomcat squadrons in the U.S. Navy. The squadron's CAG aircraft retained elements of its high-visibility markings, while the remaining aircraft were finished in the low-visibility Tactical Paint Scheme (TPS).

F-4S Phantom II (153879) of VF-151 "Vigilantes" in March 1986. This aircraft is notable for making the final operational F-4 Phantom II launch from USS Midway (CV-41) on March 24, 1986.

EA-6B Prowler (158805) of VAQ-137 "Rooks" from the September 10, 1980 to May 5, 1981 deployment aboard USS Ranger (CV-61) in the Westpac. and Indian Ocean. The Prowler was another newcomer to the fleet in the mid-1970s and soon began replacing the EKA-3B Skywarriors in the role of airborne electronic warfare.

RF-8G Crusader (144607) of VFP-63 "Eyes of the Fleet" from the August 20, 1981 to March 23, 1982 deployment aboard USS Coral Sea (CV-43) in the Indian Ocean and Westpac. The aging Crusader was still in service in the early 1980s, but not for long. TARPS-equipped F-14 Tomcats had largely replaced the Crusader in the photo-reconnaissance role; however, the Midway-class carriers (such as USS Coral Sea) were unable to operate the F-14, which meant that CVW-14 would include RF-8Gs in the air wing one last time on this deployment.

F-4N Phantom II (152981) of VMFA-531 "Grey Ghosts" from November 13, 1979 to June 11, 1980 aboard USS Coral Sea (CV-43) to the Westpac. and Indian Ocean. During Operation Eagle Claw, F-4N Phantom II aircraft of CVW-14 carried identification stripes on their outboard wings to prevent confusion with Iranian Phantom IIs. Similar markings were applied to A-6 Intruder, A-7 Corsair II, and F-14 Tomcat aircraft to ensure clear visual identification during operations.

TOMCAT - THE F-14A AND BEYOND

The 1980s represented a defining decade for the United States Navy's carrier-based fighter force, characterised by a fundamental shift from the heavy, analogue-based fighter doctrines of the Vietnam era toward a versatile, digital multirole paradigm. This transformation was necessitated by the increasing sophistication of Soviet long-range naval aviation threats and the logistical realities of maintaining carrier air wings (CVW) across a diverse fleet of supercarriers and aging mid-sized decks. At the heart of this evolution was the Grumman F-14 Tomcat, an aircraft that spent the 1980s maturing from a specialised fleet air defence platform into a multi-mission workhorse, while simultaneously being challenged and complemented by the introduction of the McDonnell Douglas F/A-18 Hornet. This decade saw the culmination of decades of aerodynamic research and the digital revolution in the cockpit, all of which reconfigured the technical baseline of naval power projection.

The Phantom's Twilight

By the time the 1980s began, the role of the dedicated fighter squadron had already undergone a radical shift. The McDonnell Douglas F-4 Phantom II, the primary backbone of the Navy's fighter force through the 1960s and 1970s, was gradually being relegated to secondary frontline and reserve roles as the F-14 Tomcat took centre stage. However, the development of the Phantom did not cease with the arrival of the Tomcat; the Service Life Extension Program (SLEP) was initiated during the 1970s and continued well into the 1980s. These upgrades were intended to maximise the utility of the remaining F-4 airframes, particularly for carriers like the USS *Midway* (CV-41) and USS *Coral Sea* (CV-43) that were not equipped to operate the F-14 Tomcat.

The Technical Evolution of the F-4N and F-4S Phantom II

The F-4S variant, which had entered squadron service in the late 1970s, represented the absolute zenith of the Phantom's design lineage. Based on the F-4J, the conversion to the S-model was an extensive modernisation effort that addressed the airframe's structural fatigue while introducing modern aerodynamic features. The most significant developmental change was the installation of leading-edge manoeuvring slats. These slats, which had been successful on the Air Force's F-4E variants, were retrofitted to the Navy's F-4S to dramatically improve low-speed handling and combat manoeuvrability.

In a dogfight, the "hard-wing" Phantoms of the previous decade often struggled with energy bleed during high-G turns. The leading-edge slats on the F-4S mitigated this by allowing the aircraft to pull higher angles of attack (AoA) while maintaining more favourable airflow over the wing's upper surface, thus tightening the turn radius significantly. Beyond aerodynamics, the F-4S featured the General Electric J79-GE-10B "smokeless" engines. Earlier variants of the J79 were notorious for their prominent black exhaust trails, which

Two F-4S Phantom II (157297 and 153900) and an A-6E Intruder (152894) on the deck of USS Midway (CV-41) in 1981. While the McDonnell Douglas F-4 Phantom II was entering the twilight of its service in the 1980s, Carrier Air Wing 5 continued to operate the aircraft until 1986, when it retired the "Spook" as the last frontline U.S. Navy carrier air wing to do so. (PHOTO: OFFICIAL US NAVY PHOTO)

provided an unintended visual beacon for enemy pilots and Ground based Anti-Aircraft defence. The J79-GE-10B utilised revised combustion liners to eliminate this visual signature, a critical development for survivability in the visual-range arena.

The SLEP initiative for the F-4S also included a comprehensive rework of the aircraft's internal "nervous system." Technicians replaced miles of aging wiring and updated the hydraulic systems to meet the higher stresses of modern carrier operations.

While the F-4S represented the most advanced iteration of the Phantom, the U.S. Navy simultaneously operated the F-4N, an upgraded version of the earlier F-4B. The development of the F-4N focused on structural reinforcement and modest avionics enhancements designed to extend the service life of the oldest Phantoms remaining in the fleet.

These aircraft were essential for Carrier Air Wing Fourteen (CVW-14) aboard the USS *Coral Sea (CV-43)*. Due to its deck configuration, the vessel was better suited for the F-4N than the heavier F-4S; unlike its sister ship, the USS *Midway* (CV-41), the *Coral Sea* had not been retrofitted with strengthened catapults. The Phantom II soldered on in several other wings aswell, but even with the upgrades, the Phantom's tenure was limited. Its role in frontline carrier air wings served primarily as a stopgap measure until the F-14 Tomcat and F/A-18 Hornet could be fully deployed from their respective production lines.

The F-14A Tomcat: The Cornerstone of Fleet Air Defence

By the early 1980s, the Grumman F-14 Tomcat had fully assumed its role as the premier air superiority fighter of the US Navy. Designed as a direct response to the failure of the General Dynamics F-111B program, the Tomcat was engineered to carry the massive Hughes AN/AWG-9 weapons control system and the long-range AIM-54 Phoenix missile. Throughout the 1980s, the F-14A underwent a series of block changes and internal modifications

F-14A Tomcat (161141) of VF-84 "Jolly Rogers" from the August 29, 1980 to October 17, 1980 deployment aboard USS Nimitz (CVN-68) during Exercise Teamwork ´80. This TARPS-capable Tomcat was flying off the deck of USS Nimitz (CVN-68) as part of the first squadron to utilise this new reconnaissance pod, which expanded the Tomcat's mission profile and, in the process, rendered dedicated photo reconnaissance aircraft such as the RF-8 and RA-5 obsolete.

F-14A Tomcat (159621) of VF-24 "Fighting Renegades" from the October 20, 1981 to May 23, 1982 deployment aboard USS Constellation (CV-64) to the Westpac. and Indian Ocean. The VF-24 "Fighting Renegades" was one of only two squadrons—alongside VF-211 "Fighting Checkmates"—to transition directly from the F-8 Crusader to the F-14 Tomcat. At the time, this transition generated some controversy, as two experienced Phantom squadrons—VF-92 and VF-96—both of which had extensive two-crew fighter experience, were disestablished. Although this decision caused some debate in the 1970s, it was largely forgotten by the 1980s, as both VF-24 and VF-211 went on to become long-serving Tomcat squadrons with distinguished operational records.

A F-14A Tomcat (161433) of VF-142 "Ghostriders" misses the wire on USS Dwight D. Eisenhower (CVN-69) on April 27, 1983 on the first day of the April 27, 1983 to December 2, 1983 deployment to the Mediterranean Sea. The aircraft in the foreground is BuNo 161429.
(PHOTO: OFFICIAL US NAVY PHOTO BY PH3 STEVEN C. HURD)

-14A Tomcat (161298) of VF-194 "Red Lightnings" operating from NAS Miramar in 1986. VF-194 and its sister squadron VF-191 "Satan's ittens" had an unusual service history. In the late 1970s, both units transitioned from the F-8 Crusader to the F-4 Phantom II. They were ubsequently disestablished, only to be reestablished in 1986 in anticipation of becoming the first squadrons to receive the F-14D Tomcat.

-14A+ Tomcat (162919) of VF-74 "Be-Devilers" is shown prior to the squadron's first deployment with the upgraded F-14A+ variant late 1989. CVW-7 was the first air wing to receive the F-14A+ variant, with CVW-17 (to which the Be-Devilers were assigned) llowing soon afterward. Many of the remaining carrier air wings saw their assigned fighter squadrons transition to the F-14A+—later edesignated the Grumman F-14B Tomcat—throughout the 1990s.

to address its initial teething problems and to integrate new technologies that were not available at its 1972 introduction.

The Engine Dilemma

Despite its aerodynamic prowess, the F-14A was hindered throughout the early 1980s by its Pratt & Whitney TF30-P-412/414 engines. Originally an interim powerplant solution, the TF30 remained the standard for the A-model for over a decade. The engine was susceptible to compressor stalls, particularly when the pilot attempted rapid throttle changes at high angles of attack or low speeds—common conditions in a dogfight. This led to a developmental focus on engine control modifications, such as the P-414A upgrade, which sought to improve the engine's reliability and resistance to foreign object damage (FOD) through reinforced fan blades.

To assist pilots in managing the TF30's delicate operating envelope, Grumman and the Navy developed the Digital Flight Control System (DFCS) in the late 1980s. The DFCS was designed to provide crisp response to pilot inputs while protecting the airframe against unrecoverable flat spins—a direct consequence of asymmetric thrust caused by a single-engine stall.

The Reconnaissance Pivot: TARPS Integration

A significant milestone occurred in 1981 with the introduction of the Tactical Airborne Reconnaissance Pod System (TARPS). As dedicated reconnaissance platforms like the Crusader, Vigilante, and Phantom were phased out, the U.S. Navy required a high-speed replacement to maintain its tactical intelligence-gathering capabilities. Given its exceptional payload capacity and speed, the F-14 Tomcat was the logical successor for this role.

While photoreconnaissance had been a staple of carrier air wings since the early 1950s, the rapid technological advancements of the 1970s and 80s shifted the focus toward multirole platforms rather than single-mission specialists. Consequently, the entire "photo recce" community was largely rendered obsolete by the introduction of a single pod.

Because the USS *Coral Sea* (CV-43) was incapable of launching the Tomcat, its embarked Carrier Air Wing (CVW-14) became the last to operate the RF-8G Crusader. Similarly, the forward-deployed USS *Midway* (CV-41) continued to operate dedicated reconnaissance aircraft well into the 1980s; however, it did so through detachments of the Marine Corps squadron VMFP-3, which utilised the RF-4B Phantom II—a variant operated exclusively by the U.S. Marines.

A F-14A Tomcat (162598) of VF-2 "Bounty Hunters" awaiting refuelling from a USAF KC-10 Extender during the February 24, 1989 to August 24, 1989 deployment aboard USS *Ranger* (CV-61).
(PHOTO: OFFICIAL US NAVY PHOTO BY LTC DAVID BARANEK)

The F-14A+

The F-14B program began when the Navy took a Tomcat prototype (BuNo 157986) out of storage and fitted it with General Electric F101DFE engines for testing in 1981. The tests were overwhelmingly successful, demonstrating that the Tomcat could accelerate from Mach 0.8 to Mach 1.8 in just 90 seconds—a massive improvement over the TF30-powered A-model.

The production version of this upgrade, the F-14A+, featured the F110-GE-400 engine. This engine produced 27,000 lbs of thrust (compared to the TF30's 20,900 lbs) and significantly improved the aircraft's rate of climb by 61%. More importantly, the F110 allowed the Tomcat to launch from carrier catapults using only "military power" (non-afterburner). This wa[s] a critical developmental achievement, as it saved thousands of pounds of fuel per launch and eliminated the massive 40-mil[e] visual signature of an afterburner flame during night operations. The squadron VF-142 and VF-143 became the first squadron[s] to replace the F-14A with the F-14A+ (B). The squadrons were a part of CVW-7 aboard USS *Dwight D. Eisenhower* (CVN-69) and took the upgraded Tomcats to se[a] for the first time on April 17, 1989, when IK[E]

eaded out on a shakedown deployment
ı the Western Atlantic.

ighter development

'he development of these fighters
ʼas also shaped by advancements in
ıaterials. While the Phantom was almost
ntirely metal, the F-14 and F/A-18 began
ɔ incorporate composite materials. The
-14 utilised graphite-epoxy composite for
s horizontal stabilisers to save weight.
he F/A-18 went further, using composites
ɔr a large percentage of its skin panels,
ʼhich provided high strength-to-weight
ıtios and increased resistance to the corrosive salt-water environment of carrier decks.

The 1980s was the decade that modernised the United States Navy's air wing, moving it from the analogue age of the Phantom into the digital age of the Super Tomcat and the Hornet. The developmental trajectory of this era was defined by the replacement of mechanical systems with digital computers, the resolution of long-standing propulsion issues through the "Great Engine War," and the expansion of aircraft utility through multirole integration. By the end of 1989, the Navy had transitioned from an air wing of specialised but limited platforms to a force of flexible, digitally-linked strike fighters. The engineering successes of this decade—from the titanium wing box of the F-14 to the quad-redundant flight controls of the F/A-18—established the technical foundation upon which naval aviation would rest for the next thirty years. The 1980s did not just produce faster and more powerful aircraft; it produced smarter aircraft, capable of adapting to a rapidly changing global threat environment through software updates and modular sensor integration, a legacy that continues to define the modern carrier deck.

INTRUDER - THE A-6E TRAM

Within the carrier air wing (CVW) ecosystem of the 1980s, the A-6E stood as the only aircraft capable of delivering heavy ordnance loads at night or in adverse weather with pinpoint accuracy. Although the introduction of the F/A-18 Hornet was intended to modernise the light-attack role, the A-6 remained indispensable due to its superior range, payload capacity, and the highly specialised coordination between its pilot and bombardier/navigator (BN). The aircraft's evolution during this decade reflected a broader naval strategy that emphasised long-range interdiction and maritime strike, using digital avionics to counter the growing sophistication of Soviet integrated air defence systems.

To understand the A-6's development in the 1980s, it is necessary to examine the organisational structure of the carrier air wing during the Reagan-era naval buildup. Under Secretary of the Navy John Lehman, the Navy adopted the "Maritime Strategy," which envisioned carrier battle groups operating in high-threat environments such as the Norwegian Sea and the North Pacific to challenge Soviet naval bastions directly. This concept required a balanced air wing composed of specialised platforms, with the A-6E Intruder serving as the primary medium-attack asset.

A typical A-6 squadron included several KA-6D tankers—converted A-6A airframes dedicated to aerial refuelling. This "organic" tanking capability was essential to sustained carrier operations, enabling strike packages to reach deep inland targets without relying on land-based Air Force support.

The A-6 entered the 1980s in its A-6E configuration, which had begun replacing the earlier A-6A and the specialised A-6B and A-6C variants developed during the Vietnam War. Although the airframe remained largely unchanged, the A-6E represented a comprehensive internal modernisation, transitioning from analogue to digital avionics.

The original A-6A was built around the Digital Integrated Attack and Navigation Equipment (DIANE) system. While revolutionary in concept, DIANE suffered from significant reliability problems, particularly in the humid climate of Southeast Asia. It employed a Norden AN/APQ-92 search radar and a separate AN/APG-46 tracking radar, both governed by drum-based computers. The A-6E replaced this fragmented architecture with the AN/APQ-148 multi-mode radar, integrating search, ground-mapping, and tracking functions into a single system.

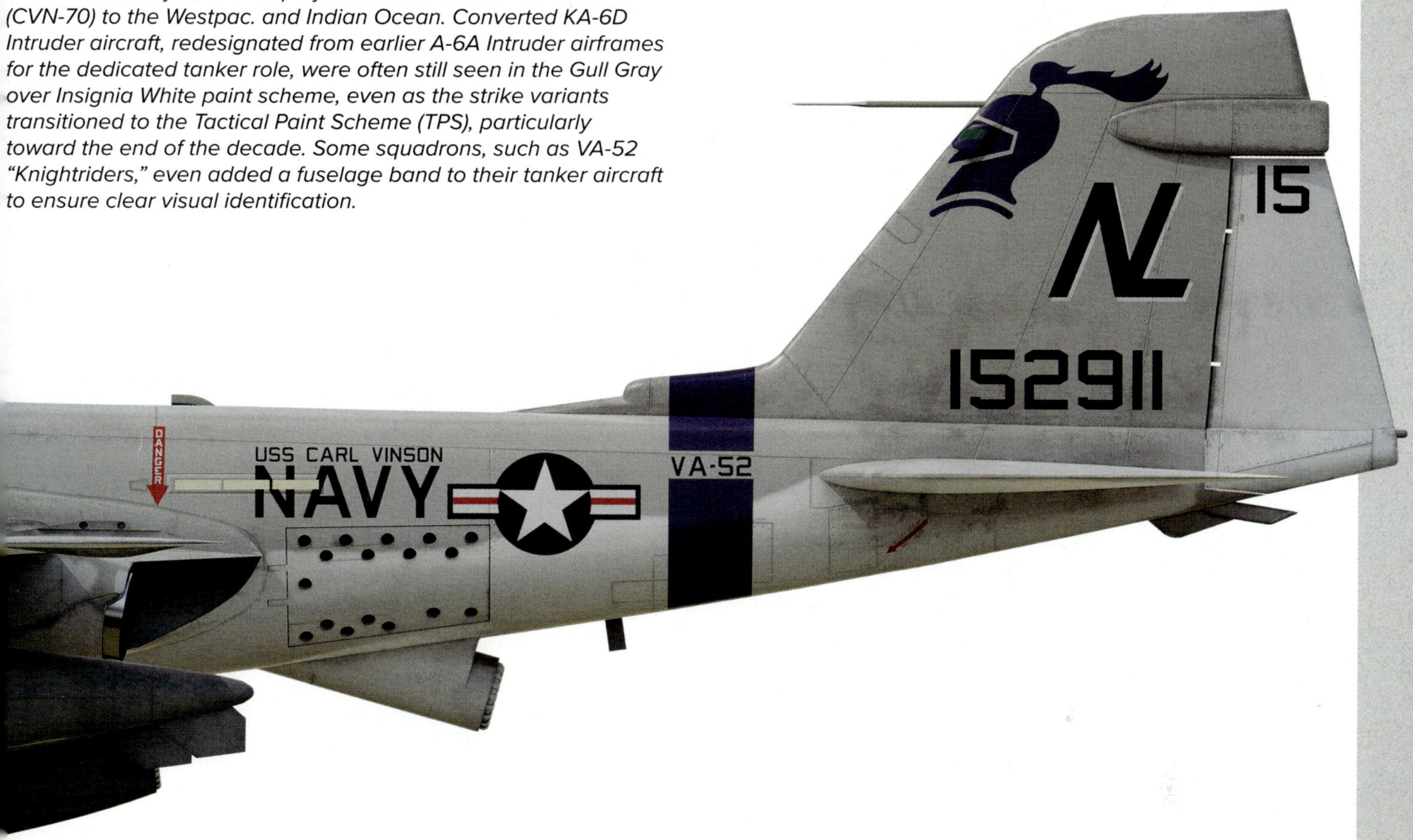

KA-6D Intruder (152911) of VA-52 "Knightriders" on the August 12, 1986 to February 5, 1987 deployment aboard USS Carl Vinson (CVN-70) to the Westpac. and Indian Ocean. Converted KA-6D Intruder aircraft, redesignated from earlier A-6A Intruder airframes for the dedicated tanker role, were often still seen in the Gull Gray over Insignia White paint scheme, even as the strike variants transitioned to the Tactical Paint Scheme (TPS), particularly toward the end of the decade. Some squadrons, such as VA-52 "Knightriders," even added a fuselage band to their tanker aircraft to ensure clear visual identification.

The A-6E also introduced the Carrier Airborne Inertial Navigation System (CAINS), specifically the AN/ASN-92, which provided substantially improved positional accuracy during long overwater flights. This was paired with the AN/ASQ-133 digital computer, enabling more efficient management of complex mission data than earlier analogue ballistic computers.

TRAM (Target Recognition Attack Multisensor)

The most significant technical advancement for the Intruder in the 1980s was the fleet-wide adoption of the Target Recognition Attack Multisensor (TRAM) system. TRAM evolved from the A-6C's TRIM (Trails/Roads Interdiction Multi-sensor) pod, used during the Vietnam War to track North Vietnamese truck traffic at night. Whereas the TRIM pod was heavy and frequently unreliable, TRAM was a fully integrated, gyro-stabilised sensor housed in a compact turret beneath the aircraft's nose.

TRAM enabled the BN to identify targets using forward-looking infrared (FLIR) imagery at night and to deliver precision-guided munitions with exceptional accuracy. Its integration with the upgraded AN/APQ-156 radar (a development of the APQ-148) allowed radar-detected targets to be handed off directly to the FLIR turret. Once the BN locked onto a target, the aircraft's computer-generated steering cues for precise ordnance release. By the mid-1980s, nearly all fleet A-6Es had been upgraded to the TRAM standard, making it the definitive configuration of the decade.

The KA-6D Tanker

In the 1980s, aerial refuelling was as vital as strike operations. The KA-6D served as the Intruder's dedicated tanker variant, created by converting older A-6A airframes into high-capacity refuelling platforms in response to a 1968 Navy requirement for an organic carrier-based tanker capable of supporting high-speed strike packages.

The conversion process was extensive. Mission avionics—including search and tracking radars and the DIANE suite—were removed. In their place, Grumman installed an internal hose-and-reel refuelling system in the rear fuselage, with a drogue assembly extending from the tail section.

Because the KA-6D frequently launched at maximum take-off weight and often recovered with substantial quantities of transferable fuel still onboard, structural reinforcements were required. These included strengthened fuselage bulkheads and modified outer wing panels to

Two A-6E Intruders (161083 and 160997) of VA-65 "Tigers" are seen in flight on April 20, 1987. At this time, CVW-7 was not deployed, but it would embark aboard USS *Dwight D. Eisenhower* (CVN-69) for a shakedown deployment in June 1987. Note that aircraft 161083 had been upgraded with the TRAM (Target Recognition and Attack Multisensor) system, while 160997 had not. This difference is visible in the distinct bulge under the nose of 161083, which is absent on 160997.
(PHOTO: OFFICIAL US NAVY PHOTO)

withstand the stresses of heavy catapult launches and arrested recoveries.

To fulfill its tanker role effectively, the aircraft's fuel system was redesigned for rapid, high-volume transfer. In addition to internal fuel, the KA-6D typically carried four or five 300-gallon external tanks on its wing and centreline pylons, greatly increasing its offload capacity.

Although its strike radar was removed, the KA-6D retained an advanced navigation suite to ensure accurate positioning during refuelling operations. Over time, upgrades included the Omega global navigation system and the ASN-41 navigational computer, improving long-range navigational precision.

To enhance mission reliability, the aircraft often carried a D-704 buddy refuelling pod on the centreline station. This provided redundancy to the internal hose-drum unit, ensuring refuelling capability even in the event of a primary system malfunction.

Operationally, the KA-6D functioned in two principal modes: recovery tanking and mission tanking. In the recovery, or "overhead," role, it orbited above the carrier to support aircraft low on fuel or

experiencing landing difficulties. In the mission role, it accompanied strike packages, providing fuel at designated waypoints to extend the range of shorter-legged aircraft such as the A-7 and F/A-18.

The Structural Challenge

As the decade progressed, the A-6 fleet faced a serious challenge unrelated to enemy action: structural fatigue. Designed before the advent of advanced fatigue-life modelling, the Intruder's airframe endured heavy ordnance loads and the repeated stresses of catapult launches and arrested landings. Over time, cracks developed in wing roots and main spars across a significant portion of the A-6E fleet.

By the mid-1980s, the Navy imposed strict flight restrictions, limiting G-loading and external store weights. This structural crisis threatened to ground the Navy's only all-weather medium-attack aircraft at a time when the Maritime Strategy demanded maximum readiness.

To extend service life, the Navy contracted Boeing Military Airplane Company to design and manufacture new wings for the A-6E. These were not simple replacements but advanced composite structures made from graphite, epoxy, and titanium. Approximately 85 percent of the fleet was slated to receive these new wings, effectively resetting the aircraft's structural lifespan and enabling continued service into the 1990s.

A-6E Intruder (159178) of VA-176 "Thunderbolts" on the October 16, 1984 to February 19, 1985 aboard USS Independence (CV-62) to the Mediterranean Sea and Indian Ocean.

A-6E Intruder (161688) of VMA(AW)-121 "Green Knights" on the February 24, 1989 to August 24, 1989 deployment aboard USS Ranger (CV-61) to the Westpac. and Indian Ocean. In the mid- to late 1980s, Carrier Air Wing 2 was often referred to as the "Grumman Air Wing," as it consisted primarily of Grumman F-14 Tomcat, Grumman A-6 Intruder, Grumman EA-6B Prowler, and Grumman E-2 Hawkeye aircraft. This concept emerged from an effort to reduce the number of different aircraft types within the air wing, thereby simplifying logistics and decreasing the variety of spare parts required to support embarked operations.

SWIP (Systems Weapons Improvement Program)

While the composite wing program addressed structural integrity, the Systems Weapons Improvement Program (SWIP), implemented in the late 1980s and early 1990s, modernised the aircraft's offensive capabilities. SWIP was an avionics and weapons-integration initiative designed to enable the A-6E to employ a new generation of standoff munitions.

The upgrade transformed the A-6E from a conventional gravity-bomb platform into a versatile missile carrier with enhanced survivability and strike flexibility.

Following SWIP, the A-6E could deploy the AGM-84 Harpoon as its primary long-range anti-ship weapon against heavily defended surface combatants. For suppression of enemy air defenses (SEAD), it could carry the AGM-88 HARM, designed to home in on hostile radar emissions. The upgrade also integrated the AGM-65 Maverick for precision strikes against mobile or hardened tactical targets.

For deep land-attack missions, the aircraft could employ the AGM-84E SLAM, a derivative of the Harpoon adapted for extended-range strikes against land targets. Additionally, it could carry the AGM-123 Skipper II, a rocket-boosted, laser-guided weapon derived from the GBU-16, combining precision guidance with increased standoff distance.

By the late 1980s, the A-6E SWIP represented one of the most capable strike platforms in the U.S. arsenal, able to carry a mixed load of anti-ship, anti-radiation, and precision land-attack weapons on a single sortie. The integration of the AN/ALR-67 radar warning receiver and the AN/ALQ-126B countermeasures system further enhanced survivability against modern surface-to-air missile threats.

Maintenance and Logistics

Sustaining the A-6E's complex avionics required a substantial logistical infrastructure. In the 1980s, oversight fell under the "REWS ON" (Reconnaissance, Electronic Warfare, Special Operations, and Naval Intelligence) program office, which managed procurement and lifecycle support for the Intruder's advanced sensors and electronic warfare systems.

Maintenance practices evolved toward more modular concepts. The introduction of BACE (Basic Automatic Checkout Equipment) enabled squadron-level technicians to diagnose faults in systems such as the APQ-148 radar or CAINS navigation unit at the intermediate maintenance level, reducing non-mission-capable time. Nevertheless, as the fleet aged and fatigue issues intensified, maintenance man-hours per flight hour remained among the highest in naval aviation—a cost considered acceptable given the aircraft's unique operational capabilities.

The Beginning of the End

By the late 1980s, continued development of the A-6 competed with funding for its intended successor, the A-12 Avenger II. Conceived as a stealthy flying wing capable of penetrating Soviet defences without extensive escort, the A-12 was expected to render the Intruder obsolete.

However, the A-12 program was beset by escalating costs and technical difficulties. Its cancellation by Secretary of Defense Dick Cheney in January 1991 left the Navy without a direct replacement for the A-6. Consequently, the service relied heavily on the A-6E SWIP and the composite wing program to sustain medium-attack capability until development of the F/A-18E/F Super Hornet could proceed. The full implications of that transition, however, belong to a subsequent chapter.

A KA-6D Intruder (BuNo Unknown) is seen from a A-6E Intruder while refuelling in May 1983.
(PHOTO: OFFICIAL US NAVY PHOTO)

A-7E Corsair II (156823) of VA-192 "Golden Dragons" from the April 14, 1981 to November 12, 1981 deployment aboard USS America (CV-66) to the Mediterranean Sea and Indian Ocean. Several A-7E aircraft from VA-192 featured a highly unconventional combination of camouflage and markings. While the airframes were finished in FS36320—the bluish-grey base of the Tactical Paint Scheme (TPS)—all markings remained in high-visibility colours. This specific aesthetic remained a rarity, appearing neither before nor after this particular deployment.

CORSAIR II AND THE HORNET — A GENERATIONAL SHIFT

The 1980s represented a watershed moment for United States naval aviation, characterised by a fundamental move away from specialised, single-mission airframes toward the era of the multirole strike fighter. This transition was crystallised in the development and fleet introduction of the McDonnell Douglas F/A-18 Hornet, an aircraft designed to replace both the McDonnell Douglas F-4 Phantom II in the fighter role and the Vought A-7 Corsair II in the light-attack role. Central to this shift was the introduction of the "VFA" designation—Strike Fighter Squadron—which officially replaced the traditional "VF" (Fighter) and "VA" (Attack) nomenclature for units transitioning to the Hornet, beginning in 1980. This chapter examines the complex engineering, developmental hurdles, and technical philosophies that defined the A-7 Corsair II and the F/A-18 Hornet throughout the 1980s, focusing exclusively on the evolutionary design processes that enabled this generational leap.

The Birth of VFA

By the late 1970s, the U.S. Navy's carrier air wings were facing a logistical and operational crisis. The F-4 Phantom II, was nearing the end of its viable operational service, and the A-7 Corsair II, while a highly efficient "bomb truck," lacked the air-to-air capabilities required for the evolving high-threat environments of the Cold War. The Navy required a platform that could streamline the carrier deck, reducing the number of different types of aircraft, spare parts, and specialised maintenance personnel required for deployment.

On November 13, 1980, the Navy instituted the VFA designation to identify squadrons equipped with the F/A-18 Hornet. This was more than a cosmetic change; it represented a doctrine of "swing-role" capability where a single aircraft could depart the carrier deck as an air-superiority escort and, with the flip of a cockpit switch, transition into a precision-strike platform. This engineering feat was enabled by advances in digital avionics and the maturation of multi-mode radar systems, which allowed the pilot to manage multiple combat roles without a second crew member.

The Navalisation of the YF-17 Cobra

The Northrop YF-17 featured a twin-engine layout and a robust structure that was more naturally compatible

A flight of three F/A-18A Hornets from VFA-25 "Fists of the Fleet" flying over the Pacific Ocean during the February 20, 1985 to August 24, 1985 deployment aboard USS Constellation (CV-64)

A-7E Corsair II (160874) of VA-93 "Ravens" from the 1982 deployments aboard USS Midway (CV-41) to the Westpac. This aircraft served with VA-93 from mid-1982 to 1986 when the squadron was disestablished.

F/A-18A Hornet (163144) of VFA-161 "Chargers" from the July 24, 1987 to August 5, 1987 deployment aboard USS Enterprise (CVN-65) to the Eastpac. The short-lived VFA-161 was assigned to the equally short-lived CVW-10 from 1986 to 1988. The "Chargers" had previously served for many years with CVW-5 as a forward-deployed squadron operating the McDonnell Douglas F-4 Phantom II, making them one of the last U.S. Navy units to fly the Phantom in frontline service. After redesignation and transition to the Hornet with CVW-10, the squadron was decommissioned shortly thereafter.

F/A-18A Hornet (161960) of VFA-25 "Fists of the Fleet" from the February 20, 1985 to August 24, 1985 deployment aboard USS Constellation (CV-64) to the Westpac. and Indian Ocean. VFA-25 with VFA-113 were the first squadrons to operational deploy the F/A-18 Hornet aboard a US Navy aircraft carrier.

A-7E Corsair II aircraft from VA-146 "Blue Diamonds" and VA-147 "Argonauts" are seen on the deck of USS *Kitty Hawk* (CV-63) on March 12, 1985, during carrier qualifications (CarQuals) for the upcoming deployment to the Westpac. and the Indian Ocean. During the 1980s, it was standard for a carrier air wing to include two A-7 squadrons, and CVW-9 was no exception. Both squadrons remained part of CVW-9 until 2009, although they transitioned from the A-7 Corsair II to the F/A-18 Hornet in 1989, becoming VFA-146 and VFA-147. (PHOTO: OFFICIAL US NAVY PHOTO BY PH3 BREWER/PH3 HENSLEY)

with naval requirements, than the F-16 competition. McDonnell Douglas, as the lead contractor for the naval version, and Northrop embarked on an extensive redesign that transformed the YF-17 technology demonstrator into the F/A-18A Hornet. Although the general configuration was retained, the resulting aircraft was entirely new.

The navalisation process focused on structural reinforcement. The airframe, undercarriage, and tailhook were strengthened for the immense longitudinal and vertical loads of carrier ops. To improve low-speed lift and control, the wing area was expanded by 14%, and the leading-edge root extensions (LERXes) were redesigned. Most of the YF-17's LERX slots were filled to reduce drag, though a single slot behind the intake was retained for boundary layer air management. A critical engineering addition was the "snag" on the wing's leading edge and stabilators, introduced to prevent aeroelastic flutter that had been discovered during the development of the F-15.

A defining element of the Hornet's development was the General Electric F404-GE-400 afterburning turbofan. The F404 was a direct evolution of the YJ101, but it was refined into a "pilot's engine" that prioritised reliability, maintainability, and unrestricted operability. The developmental goals for the F404 were to provide twice the pressure ratio of the J79 while weighing half as much and having one-third fewer parts.

The F404's most significant contribution to the Hornet's multirole mission was its unrestricted throttle movement. Unlike the P&W TF30 engines found in the A-7 and F-14, which were prone to compressor stalls during rapid throttle transients or high-alpha manoeuvres, the F404 used a Digital Electronic Control (DEC) system to manage engine parameters. This allowed a pilot to "slam" the throttles from idle to full afterburner at any speed or altitude without risk of flameout.

The engine's modularity was a revolutionary development in naval engineering. The F404 consists of six modules—the fan, compressor, combustor, high-pressure turbine, low-pressure turbine, and afterburner—each of which is fully interchangeable. This design allows maintenance crews aboard carriers to swap out damaged modules without needing to replace the entire engine, drastically increasing the mission-capable rates of the Hornet compared to the A-7.

The Digital Architecture

The F/A-18 was the first Navy aircraft to incorporate a digital multiplexing avionics bus, allowing various subsystems to communicate via a centralised data processor. This architecture centred on the Hughes AN/APG-65 radar, a system that effectively enabled the merger of the F-18 and A-18 programs.

The APG-65 was the first radar to incorporate a single programmable digital signal processor (PSP) capable of both air-to-air and air-to-ground operations. In the 1970s and early 1980s, this was a radical engineering achievement. The radar could process up to 7.2 million operations per second, managing complex waveforms that previously required dedicated hardware for each mode.

Key developmental modes included: Velocity Search is a high pulse repetition frequency (PRF) radar mode designed to detect high-speed, nose-aspect targets at maximum range. By focusing on closing velocity rather than precise positional data, it allows aircrews to identify and engage fast-moving threats early, making it especially valuable in beyond-visual-range combat scenarios.

Track-While-Scan (TWS) enables the radar to maintain track files on up to ten targets simultaneously while continuing to scan the surrounding airspace for additional threats. This capability provides strong situational awareness without sacrificing search coverage.

Doppler Beam Sharpened (DBS) mode enhances the radar's air-to-ground performance. Through digital signal processing, the radar effectively "sharpens" its beam, producing high-resolution ground maps. This allows pilots to distinguish specific structures, vehicles, or ship types in virtually any weather condition, significantly improving strike accuracy and target identification.

Air-to-Surface Ranging mode supplies precise distance measurements to the mission computer, enabling highly accurate delivery of unguided "iron bombs." By providing reliable slant-range data, this mode achieves a level of bombing precision that rivals dedicated attack aircraft such as the A-7E, enhancing the platform's effectiveness in the strike role.

The radar's physical design also reflected the Navy's maintenance

An A-7E Corsair II (158664) of VA-56 "Champions" is seen on the deck of USS *Midway* (CV-41) on May 9, 1984. CVW-5, a forward-deployed air wing operating in the Westpac., was homeported at Naval Air Facility Atsugi. Its two Corsair II squadrons, VA-56 and VA-93, had a long history of operating the aircraft, but both transitioned from the A-7 Corsair II to the F/A-18 Hornet in 1986.
(PHOTO: OFFICIAL US NAVY PHOTO VIA PH1 PAUL SOUTAR)

07
NF
NAVY

requirements. It was housed in a slide-out nose rack, allowing technicians to pull the unit forward for inspection without removing external panels or interfering with the cockpit.

A-7E Corsair II Development

As the Hornet was entering service, the A-7E Corsair II was the standard against which its strike capability was measured. The A-7E was the final fleet version of the "SLUF" (Short Little Ugly Fellow), and by the early 1980s, it had reached the pinnacle of subsonic attack engineering.

The A-7E was powered by the Allison TF41-A-2 turbofan, a non-afterburning engine that provided 15,000 lbf of thrust. While it lacked the supersonic dash capability of the Hornet, the TF41 offered exceptional fuel economy. This efficiency allowed the A-7E to maintain a combat radius of 608 NM on internal fuel, significantly outperforming the Hornet's early developmental figures. The A-7E's endurance was further aided by its ability to carry up to 15,000 pounds of ordnance on eight store stations, earning it the moniker "bomb truck".

In the 1980s, the A-7E underwent several critical avionics updates to maintain its relevance. The most notable was the introduction of the Low-Altitude Night Attack (LANA) system. This suite integrated a Forward-Looking Infrared (FLIR) pod with the aircraft's AN/APQ-126 radar and Head-Up Display (HUD), allowing the pilot to fly low-level, terrain-following missions in total darkness. The A-7E's HUD was one of the first to project a digital "flight path marker" and predictive impact point (CCIP) for unguided bombs, a technology that would be further refined in the Hornet's glass cockpit.

Transition Challenges: Radius and Bring-back Weight

The transition from the A-7E to the F/A-18A was not without controversy, primarily regarding the Hornet's performance in the attack role. Engineering compromises required to achieve supersonic multirole capability led to significant reductions in combat endurance and payload recovery compared to the specialized A-7.

One of the most criticised aspects of the Hornet's development was its combat radius. The A-7E's 608 NM radius allowed the carrier to stand off at a safer distance from the target. The Hornet, hampered by the weight of its twin engines and smaller internal fuel volume, initially struggled to meet the Navy's 400 NM threshold. After extensive testing and drag-reduction efforts, the Navy eventually accepted a best range of 370–380 NM, representing a substantial loss in loiter time and penetration depth.

The fuel consumption rate of the F404 at subsonic cruise was roughly 50% higher than that of the TF41, a consequence of the bypass ratio chosen for supersonic performance. This "short-legged" nature of the Hornet meant that the Navy had to continue relying on the Grumman A-6E Intruder for long-range, all-weather strike, and it eventually drove the development of the "Bomb Cat" strike modifications for the F-14 Tomcat.

The Carrier Recovery Weight Limitation

A critical developmental hurdle for the F/A-18A was its "bring-back" weight—the maximum weight of the aircraft, remaining fuel, and unexpended ordnance allowed for an arrested landing. The structural limits of the F/A-18A's arresting hook and landing gear meant that pilots returning to the carrier with heavy, unexpended precision-guided munitions—such as the AGM-84 Harpoon or the AGM-62 Walleye—often had to jettison these expensive weapons into the sea.

The A-7E, despite its smaller size, was engineered with a robust wing and landing gear that allowed it to recover a larger percentage of its maximum payload.29 This discrepancy remained a point of friction until the development of the F/A-18C/D in the late 1980s, which introduced further structural strengthening and improved flight control software to manage higher recovery weights.

F/A-18A Hornet (162861) of VFA-137 "Kestrels" from the September 28, 1987 to March 29, 1988 deployment aboard USS Coral Sea (CV-43) to the Mediterranean Sea. The VFA-137 "Kestrels" were among the squadrons established in the 1980s specifically to operate the F/A-18 Hornet. In contrast, most other Hornet squadrons had longer lineages, having previously served either as fighter squadrons or—more commonly—as attack squadrons before transitioning to the Hornet.

)rdnance Integration and Survivability

'hroughout the 1980s, the Hornet's levelopment focused on integrating a vide array of advanced weaponry that vould allow it to succeed in the multirole environment. The aircraft was designed vith nine external stores stations, including wo wingtip rails for AIM-9 Sidewinders, our underwing pylons, two nacelle-nounted stations for AIM-7 Sparrows or ensor pods, and a centreline pylon.

ᐊdvanced Weaponry Development

ntegration of the following systems was nalised in the mid-to-late 1980s. The ᐊGM-88 HARM, the High-Speed Anti-Radiation Missile, became a cornerstone f the Hornet's SEAD capability. The Hornet's mission computers were niquely capable of integrating with the HARM Command Launch Computer (CLC), llowing for highly automated "self-protect" or "target-of-opportunity" modes gainst enemy radars.

The AGM-84 Harpoon integration equired significant software development o enable the AN/APG-65 radar to pass argeting coordinates directly to the nissile's seeker before launch, enhancing ne aircraft's anti-ship strike effectiveness.

The transition to the F/A-18C in 1987 rought the capability to carry the AIM-120 .MRAAM, providing the Navy with its first re-and-forget, medium-range air-to-air capability outside of the F-14's Phoenix missile system.

Electronic Warfare and Survivability

To ensure survivability in the dense air defence environments of the 1980s, the Hornet shared several defensive systems with the F-14 and late-model F-4s, including the AN/ALR-50 Radar Warning Receiver and the AN/ALQ-126 electronic countermeasures jammer. This commonality allowed for streamlined training for electronic warfare officers and ensured that the Hornet could defend itself against the burgeoning threat of Soviet-made surface-to-air missiles (SAMs).

The F/A-18C/D Upgrades

Starting in 1987, production shifted to the F/A-18C and its two-seat trainer/strike derivative, the F/A-18D. While physically very similar to the "A" and "B" models, the "C" model featured massive internal upgrades that solidified the Hornet's position as the Navy's premier multirole platform.

Lot 12 Hornets and later variants introduced a significant Night Attack Capability, featuring a redesigned cockpit compatible with Night Vision Goggles (NVG) and a digital moving map system that replaced the older film-based display. These improvements greatly enhanced situational awareness and operational effectiveness during low-light and nighttime missions.

In addition, propulsion was improved with the introduction of the F404-GE-402 EPE (Enhanced Performance Engine), which produced approximately 17,700 pounds of thrust. This marked a substantial increase over the baseline engine and helped offset the additional weight from newly integrated avionics systems.

Pilot safety was also enhanced with the adoption of the SJU-17/A NACES (Navy Aircrew Common Ejection Seat), providing improved escape capability and greater protection across the aircraft's entire flight envelope.

The Multi-Mission Concept

The development and transition of the A-7 Corsair II and the F/A-18 Hornet during the 1980s fundamentally reshaped the U.S. Navy. While the A-7E remained an unmatched subsonic strike platform for sheer endurance and bomb-carrying efficiency, its retirement was necessitated by the need for a more versatile, self-defending airframe. The F/A-18 Hornet, despite early criticisms regarding its range and recovery weight, proved that a single aircraft could successfully integrate the roles of fighter and attack through the use of digital avionics, a programmable multimode radar, and a highly reliable twin-engine powerplant.

Two F/A-18A Hornets of VFA-192 "Golden Dragons" are seen in flight in June 1989. The squadron was part of Carrier Air Wing 5 (CVW-5), which had been deployed aboard USS *Midway* (CV-41) since 1973. This particular deployment, lasting from May 31 to July 25, 1989, took the air wing to the Westpac.
(PHOTO: OFFICIAL US NAVY PHOTO BY LT.CDR T.B. SURBRIDGE, USN)

USS MIDWAY
313
M
MIDWAY
CDR J.A. PETTITT
"GRUMP"
301
M

An EA-6B Prowler (161348) of VAQ-138 "Yellow Jackets" is seen taxiing at NAS Fallon prior to take-off in April 1986. The squadron would later embark aboard USS *Nimitz* (CVN-68) with Carrier Air Wing 9 (CVW-9) for a deployment to the Norlant, which lasted from August 15 to October 16, 1986.
(PHOTO: OFFICIAL US NAVY PHOTO BY MICHAEL GROVE)

EA-6B PROWLER AND ELECTRONIC WARFARE

The technological landscape of the 1980s necessitated a fundamental reassessment of the United States Navy's electronic warfare (EW) capabilities. While the Grumman EA-6B Prowler had entered service in the early 1970s as a replacement for the EKA-3B Skywarrior, the rapid advancement of Soviet-bloc integrated air defence systems (IADS) during the late Cold War era demanded a shift from manual, frequency-limited systems to automated, high-speed architectures. The development of the Prowler in the 1980s was not merely a series of incremental updates but a comprehensive transformation that redefined the aircraft's role from a defensive escort to a multi-mission platform capable of both soft-kill electronic suppression and hard-kill kinetic destruction.

The transition into the 1980s was marked by a shift in the philosophy of the electromagnetic battle. Earlier iterations of the Prowler, such as the Standard and Expanded Capability (EXCAP) versions, relied heavily on the manual coordination of the three Electronic Countermeasures Officers (ECMOs) to identify, prioritise, and jam signals. The complexity of the 1980s threat environment—characterised by frequency-agile radars and low-probability-of-intercept (LPI) techniques—rendered these manual processes insufficient. Consequently, the developmental efforts of this decade focused on the integration of advanced digital computing, high-speed data buses, and software-driven signal generation.

Structural and Aerodynamic Architecture: The Platform Foundation

Derived from the A-6 Intruder, the Prowler featured a 4.5-foot fuselage extension to accommodate a four-seat cockpit. This structural stretch was essential for housing the specialised mission equipment and the crew required to manage it. To support the massive increase in electronic components and the carriage of up to five 950-pound AN/ALQ-99 jamming pods, the airframe underwent extensive reinforcement of the wing centre section and the landing gear.

The ICAP II

The introduction of the Improved Capability II (ICAP II) configuration in 1980 represented the most significant technological leap for the Prowler during this decade. The ICAP II prototype first flew on June 24, 1980, ushering in an era where software would dictate the success of the electronic attack mission. The core of this upgrade was the transition from the legacy AN/AYK-6 weapons control computer to the modern, modular AN/AYK-14(V) Tactical Computer.

The AN/AYK-14 was a 16-bit general-purpose computer that utilised large-scale integration (LSI) components to reduce parts count while increasing processing throughput. Its modular architecture, composed of 6-by-9-inch printed wiring boards, allowed the Navy to configure the system for specific requirements, including expanded memory modules and high-speed input/output (I/O) channels. In the EA-6B, the AYK-14 functioned as the central nervous system, interfacing with the AN/ALQ-99 Tactical Jamming System.

This digital backbone allowed for a redistribution of the workload among the crew. In earlier versions, the ECMO 1 in the front seat was overtaxed with communication and navigation duties while attempting to manage electronic warfare tasks. The ICAP II configuration moved the communication jamming functions to the ECMO 1 position, while the two rear ECMOs (positions 2 and 3) focused on the radar jamming mission. This shift was supported by the introduction of the ASN-123 signal data converter and digital display group, which provided the crew with clearer, computer-generated threat information.

The Evolution of the AN/ALQ-99

Throughout the 1980s, the AN/ALQ-99 Tactical Jamming System (TJS) remained the primary offensive tool of the Prowler. The ICAP II upgrade introduced the AN/ALQ-99D configuration, which featured major improvements to the external jamming pods. Each pod was a self-contained unit housing two high-power transmitters, a control computer, and an exciter. A ram air turbine (RAT) at the front of each pod provided independent electrical power, ensuring that the heavy power requirements of the jammers did not drain the aircraft's internal electrical systems.

The most groundbreaking development for the ALQ-99 in the 1980s was the "Universal Exciter". In earlier iterations, jamming pods were configured for specific, single frequency bands and could not be altered once the aircraft was in flight. The Universal Exciter, however, was totally programmable, allowing the system to adjust its parameters based on the specific threats detected during a mission. This provided the ability for each pod to generate signals in any one of seven frequency bands and to jam two different frequency bands simultaneously.

The jamming techniques were also improved through computer-managed power allocation. The AYK-14 computer allowed for "Power Management," a technique that prioritised the greatest threats identified by the ALR-42 System Integration Receiver (SIR) and focused the maximum available jamming energy on

EA-6B Prowler (160791) of VAQ-130 "Zappers" on the June 7, 1982 to November 16, 1982 deployment aboard USS Forrestal (CV-59) to the Mediterranean Sea and the Indian Ocean. The EA-6B Prowler became the mainstay of U.S. Navy electronic warfare operations after taking over the mission from the EKA-3B Skywarrior. Representing a significant technological advancement in a much more compact airframe, the Prowler was particularly well suited to its intended role. It remained an essential component of carrier air wings throughout the decade and a key asset in the U.S. Navy's electronic warfare capabilities.

EA-6B Prowler (161243) of VAQ-132 "Scorpions" on the April 27, 1983 to December 2, 1983 deployment aboard USS Dwight D. Eisenhower (CVN-69) to the Mediterranean Sea. The transition to the Tactical Paint Scheme (TPS) was well underway by the early 1980s. Although standard colours and patterns had been defined, paint suppliers were not always able to deliver consistent shades and hues. As a result, field repairs and repainting often produced noticeable variations, and this aircraft provides a good example of how such maintenance work could result in a uniquely varied camouflage scheme.

those targets. This was facilitated by beam steering technology in the transmitters, which could focus the jamming signal precisely at the threat rather than radiating energy in all directions.

The AGM-88 HARM Integration

While the Prowler was initially conceived as a non-lethal suppression platform, the tactical realities of the 1980s demanded a "hard-kill" capability. The integration of the AGM-88 High-speed Anti-Radiation Missile (HARM) was a watershed moment that transformed the EA-6B into a dual-threat platform. The HARM was designed to home in on enemy radar emissions and destroy the transmitter and antenna, providing a permanent solution to air defence threats.

The HARM integration began with the ICAP II program and was retrofitted to earlier aircraft. The missile was a large, high-performance weapon, weighing nearly 800 pounds and capable of speeds exceeding Mach 2. Its speed was a critical factor, as it minimised the time an enemy radar operator had to shut down the transmitter upon detecting a launch. The Prowler typically carried up to four HARMs on its underwing pylons, though a mission loadout often mixed missiles with jamming pods and external fuel tanks.

The Advanced Capability (ADVCAP) Initiative

In the late 1980s, the Navy recognised that the aging EA-6B airframe was reaching its aerodynamic and weight limits. This led to the Advanced Capability (ADVCAP) program, the most ambitious modernisation effort in the Prowler's history. ADVCAP was designed to address three areas: avionics, propulsion, and aerodynamics. The first flight of the upgraded Prowlers took place on 29 October 1989 just as the decade ended.

The aerodynamic portion, known as the Vehicle Enhancement Program (VEP), was a direct response to the Prowler's deteriorating handling qualities at high gross weights. The VEP modifications included leading-edge strakes at the wing-fuselage junction to improve stability at high angles of attack, a vertical fin pod extension to increase rudder effectiveness, and new digital flight controls (SAFCS) to assist the pilot in the critical landing phase.

The ADVCAP avionics suite centred on the AN/ALQ-149 and an improved receiver set, providing a "glass cockpit" for all four crew members. While the ADVCAP program was ultimately cancelled in 1995 due to post-Cold War budget cuts, it produced three flying prototypes including the famous "FrankenProwler" BuNo 158542.

An EA-6B Prowler (161881) of VAQ-137 "Rooks" is seen seconds before launch from the deck of USS *Saratoga* (CV-60) on January 28, 1986, during a deployment that lasted from August 26, 1985, to April 16, 1986. The aircraft is equipped with three AN/ALQ-99 jamming pods and two drop tanks, a typical loadout for this deployment, during which CVW-17 operated in the Mediterranean Sea amid the standoff with Libya that ultimately culminated in Operation El Dorado Canyon.
(PHOTO: OFFICIAL US NAVY PHOTO BY PH1 WILLIAM A. SHAYKA)

606

E-2C Hawkeye (161097) of VAW-123 "Screwtops" from the March 10, 1986 to September 10, 1986 aboard USS America (CV-66) to the Mediterranean Sea. The Grumman E-2 Hawkeye was an essential component of every carrier air wing in the 1980s, providing airborne early warning and navigation support for aircraft operating both within the air wing and in the surrounding operational area.

VIKINGS, HAWKEYES, AND HELICOPTERS

The decade of the 1980s represents the pivotal era in which the United States Navy shifted from post-Vietnam retrenchment to an assertive, forward-oriented maritime strategy that demanded unprecedented technological integration among carrier-based platforms. Defined by the "600-ship Navy" initiative and the doctrine of the "Outer Air Battle," this period emphasised intercepting Soviet long-range aviation and submarine threats before they could launch standoff weapons against carrier battle groups. Central to this transformation were three aircraft: the E-2C Hawkeye, the S-3B Viking, and the SH-60 Seahawk. Over the course of the decade, these platforms evolved from relatively specialised sensor aircraft into networked components of an increasingly digital battlespace. Their development was not merely incremental improvement in performance, but a fundamental digitalisation of naval warfare, characterised by expanded computing capacity, advanced signal processing, and the integration of precision-guided standoff weapons.

The E-2C Hawkeye: Maturation of the Airborne Command and Control Node

By the early 1980s, the E-2C Hawkeye had established itself as the indispensable airborne early warning and command-and-control platform of the carrier force. Compared to the earlier E-2A and E-2B variants, the E-2C represented a major advance, particularly through the introduction of the AN/APS-125 radar and fully digitalised mission systems. A critical innovation of the period was the addition of the Airborne Moving Target Indicator (AMTI), which enabled detection and tracking of moving ground targets. This capability expanded the aircraft's function beyond maritime surveillance to include coordination of overland strike operations and close air support.

Radar Evolution: From APS-125 to the Group 2 Standard

The evolution of the E-2C during the 1980s is most clearly reflected in its radar development. Early "Basic Charlie" aircraft were equipped with the AN/APS-125, capable of tracking approximately 300 targets at ranges approaching 290 miles

S-3A Viking (160163) of VS-21 "Redtails" from the January 5, 1988 to July 3, 1988 deployment aboard USS Enterprise (CVN-65) to the Westpac., Indian Ocean and Persian Gulf. The S-3A variant was a common sight in all air wings aboard the US Navy's super carriers throughout the decade.

within a surveillance volume of roughly three million cubic miles. As Soviet electronic warfare capabilities advanced, the Navy introduced the "Group 0" configuration beginning in 1980.

Although initially retaining the APS-125 designation, the upgraded system was redesignated AN/APS-138 by 1984. Its most significant enhancement was the Total Radiation Aperture Control-Antenna (TRAC-A), which reduced sidelobe emissions and limited the vulnerability of the radar to enemy electronic intelligence collection and jamming. This improvement enhanced performance in dense electromagnetic environments and countered the sophisticated electronic warfare techniques associated with Soviet Tu-22M Backfire bombers and related platforms.

The subsequent Group 1 configuration introduced the AN/APS-139 radar, supported by a more powerful Litton processor that increased track capacity dramatically—from roughly 600 to 2,400 targets. In 1988, the AN/APS-145 entered service, providing a 40 percent increase in detection range and nearly doubling surveillance volume. Its advanced electronic counter-countermeasures and adaptive signal processing enabled effective operation in complex terrain and heavy jamming conditions, preparing the platform for continued service into the 1990s.

The 48KB Revolution and Mission Computing

A defining illustration of 1980s technological progress was the increase in mission computer memory from 16 kilobytes to 48 kilobytes in the Group 0 aircraft. Though modest by modern standards, this threefold expansion represented a substantial advance, enabling the processing demands of the APS-138 radar and enhanced electronic support measures suite.

The Litton L-304 digital computer integrated radar, Identification Friend or

An E-2C Hawkeye (161344) of VAW-116 "Sun Kings" is seen parked at NAS Fallon on September 23, 1988. As indicated by its tail code, the aircraft was assigned to CVW-2, which would deploy approximately six months later, beginning on February 24, 1989. That deployment lasted until August 24, 1989, and would be the air wing's final deployment of the decade.
PHOTO: OFFICIAL US NAVY PHOTO BY MICHAEL GROVE)

S-3A Viking (160144) of VS-24 "Scouts" from the December 30, 1988 to June 30, 1989 deployment aboard USS Theodore Roosevelt (CVN-71) to the Mediterranean Sea. Aircraft of the 1980s were often designed to fulfil multiple mission roles, as illustrated here by the buddy refuelling pod mounted on the wing station of this S-3 Viking. Similar pods could also be carried by several other aircraft within the air wing, providing greater operational flexibility and extending the range and endurance of carrier-based strike packages.

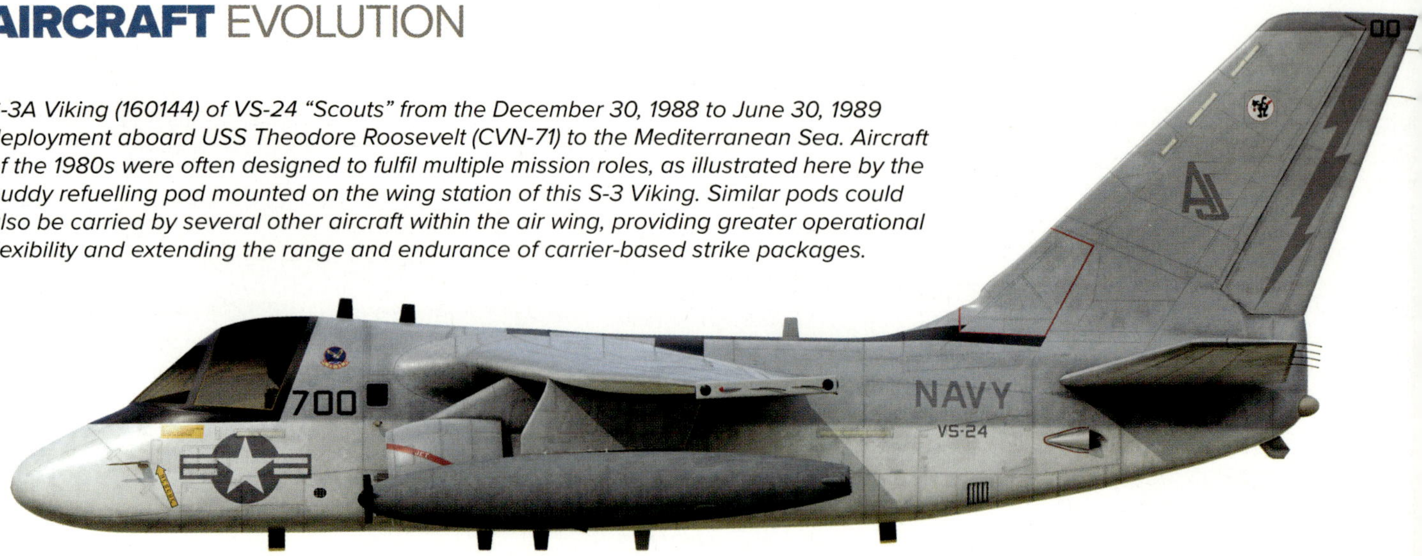

S-3B Viking (159732) of VS-31 "Topcats" from the Solid Shield 1989 exercise from April 17, 1989 to May 20, 1989 aboard USS Dwight D. Eisenhower (CVN-69) operating off the coast of South Carolina. The Lockheed S-3B Viking entered frontline carrier air wing service in the late 1980s. While the external differences compared with the earlier Lockheed S-3A Viking were relatively subtle, the updated variant could be readily identified by its modified wingtip pods.

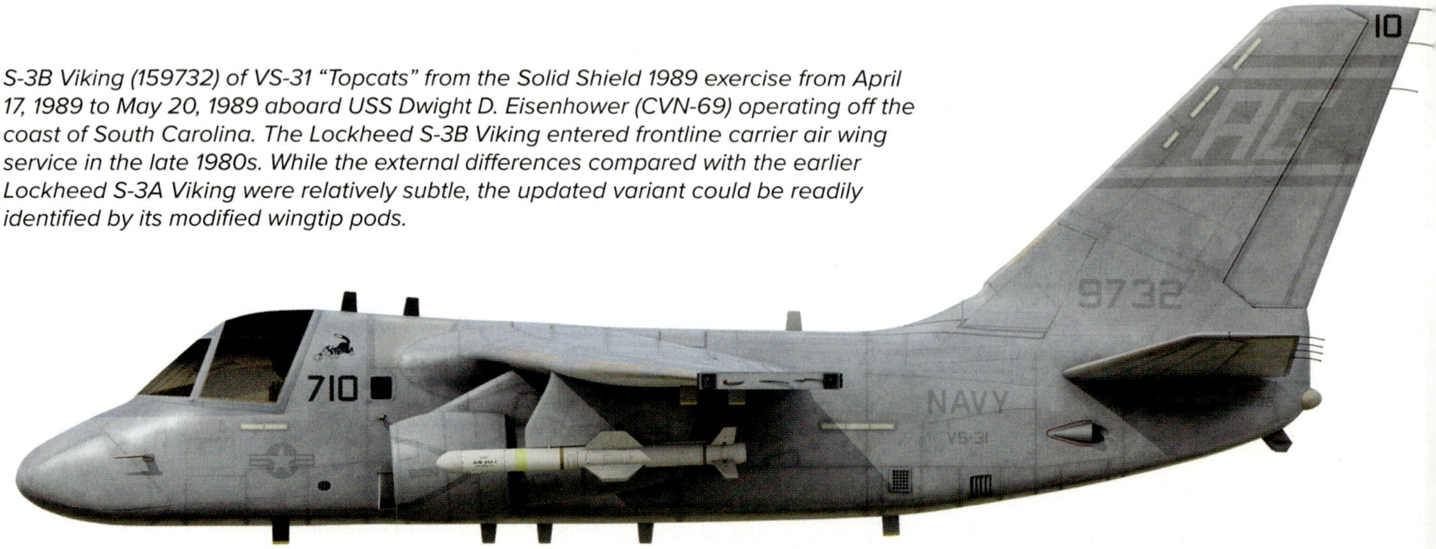

Two S-3A Vikings (BuNos unknown) of VS-31 "Topcats" are seen in flight in November 1983 during a deployment to the Mediterranean Sea that lasted from April 27 to December 3, 1983. Although these aircraft still carry the high-visibility paint scheme, their squadron-specific markings are minimal, consisting only of blue trim on the vertical stabiliser caps and the squadron insignia positioned behind the canopy. (PHOTO: OFFICIAL US NAVY PHOTO)

An S-3A Viking (BuNo unknown) of VS-32 "Maulers" is being positioned on one of the steam catapults aboard USS Theodore Roosevelt (CVN-71) on January 23, 1987. This occurred during the ship's maiden deployment, a short shakedown cruise that lasted from January 4 to February 17, 1987. Note the aircraft's more subdued color scheme as the 1980s drew to a close. Shortly thereafter, S-3 Vikings would be repainted in the Tactical Paint Scheme (TPS). (PHOTO: OFFICIAL US NAVY PHOTO BY PH3 BORBELY)

Foe (IFF), and passive sensor data into a coherent tactical display. Expanded memory allowed more sophisticated algorithms within the Airborne Tactical Data System (ATDS), improving multi-sensor track correlation and stabilising the tactical picture presented to the Carrier Air Wing. By the end of the decade, however, the limitations of these legacy processors prompted development of the Mission Computer Upgrade (MCU) program, which sought to replace older systems with commercial off-the-shelf processors such as the DEC 2100, offering far greater performance, reduced weight, and improved cooling efficiency.

Passive Detection

An equally significant development was the upgrade of the Hawkeye's passive detection systems. The transition from the AN/ALR-59 to the AN/ALR-73 Passive Detection System in the early 1980s markedly enhanced electronic support measures capability. The ALR-73 employed 52 antennas distributed across the airframe to provide full 360-degree coverage.

The Eye in the Sky: This E-2C Hawkeye (161094) of VAW-124 "Bear Aces" is seen in flight in June 1980, prior to Exercise Teamwork '80, during which it operated from USS Nimitz (CVN-68) from August 29 to October 17, 1980. The Hawkeye's role as an airborne early warning platform rarely attracted media or filmmakers' attention, yet it was crucial to carrier operations in the 1980s—a fact demonstrated by numerous examples, some of which are discussed in subsequent sections of this book. (PHOTO: OFFICIAL US NAVY PHOTO)

C-2A Greyhound (162152) of VR-24 "Lifting Eagles" based at NAS Sigonella in 1988.Sharing several components with the E-2 Hawkeye, the C-2 Greyhound featured a deeper and wider fuselage, a redesigned tail section, and a rear cargo ramp—improvements that made it a significant advancement over its predecessor, the C-1A Trader. Another important advantage was its turboprop engines, which used the same jet fuel as other aircraft in the carrier air wing, whereas the C-1 Trader's radial engines required a specialized aviation gasoline.

Using a superheterodyne receiver and advanced signal processing, the system analysed pulse characteristics, frequencies, and scan rates to identify specific emitters, whether aircraft or missile systems. Direction-finding accuracy improved from approximately five degrees to two degrees, allowing earlier and more precise threat detection. This enhanced passive capability added a critical early-warning layer in the electronic warfare environment of the Cold War.

From Sub-Hunter to Sea Control Multi-Mission Platform

During the 1980s, the Lockheed S-3 Viking underwent significant mission expansion. While its initial transition from the S-2 Tracker to the S-3A occurred in the 1970s, the decade was defined by the Weapon System Improvement Program (WSIP), which transformed the S-3A into the multi-mission S-3B. Anti-Submarine Warfare squadrons were fully integrated into carrier air wings, typically operating about ten aircraft per squadron.

The S-3B Upgrade and the Rise of Anti-Surface Warfare

The S-3B variant first flew on September 13, 1984, with operational deliveries beginning in 1987. This upgrade shifted the Viking's mission from a primarily defensive anti-submarine platform to a broader sea-control aircraft with offensive anti-surface capability.

The most visible change was integration of the AGM-84 Harpoon anti-ship missile. This addition provided a standoff strike capability against Soviet surface combatants, extending the reach of the carrier air wing beyond the range of its surface-launched weapons. The upgrade reflected the growth of the Soviet blue-water navy and the requirement for carrier aviation to project power while maintaining an anti-submarine screen.

Advanced Sensors

The S-3B's sensor suite was comprehensively improved. The legacy AN/APS-116 radar was replaced by the AN/APS-137(V)1 Inverse Synthetic Aperture Radar (ISAR). ISAR generated high-resolution images by exploiting the motion of target vessels, allowing crews to classify ships by type at standoff range, even in darkness or poor visibility.

US-3A Viking (157995) of VRC-50 "Foo Dogs" in 1987. One of only six US-3A Vikings ever to be operated by the US Navy. Like the C-1A Trader, the US-3A Viking was derived from an anti-submarine warfare aircraft. The US-3A operated from Westpac. bases such as NAF Atsugi, NAS Cubi Point, and Andersen Air Force Base. Across the vast distances of the Pacific Ocean, its greater range compared with the C-2 Greyhound proved to be a significant operational advantage.

SH-3H Sea King (149902) of HS-4 "Black Knights" from the September 18, 1989 to November 8, 1989 deployment aboard USS Carl Vinson (CVN-70) to the Norpac. and Westpac. Approaching the twilight of its career, the SH-3 Sea King nevertheless remained the primary anti-submarine warfare (ASW) and search-and-rescue (SAR) platform for U.S. Navy carrier air wings throughout the 1980s

In the acoustic domain, the AN/UYS-1 Proteus signal processor enhanced analysis of sonobuoy data, addressing the challenge posed by increasingly quiet Soviet submarines. Working with the improved AN/ARR-78 receiver, the system enabled monitoring of multiple buoys with greater discrimination and fidelity. These capabilities strengthened both anti-submarine warfare effectiveness and search-and-rescue operations.

Computing and Tactical Integration

At the centre of the S-3B's tactical capability was the AN/AYK-10 digital mission computer. Derived from systems used in the P-3C Orion, it integrated sensor inputs into a shared tactical picture for the four-man crew. The tactical coordinator could compare real-time data against stored acoustic and radar signatures, accelerating identification and decision-making. This degree of onboard integration allowed the S-3B to function not only as a sub-hunter, but also as a distributed command-and-control asset within the fleet's outer defensive zone.

Collectively, the developments of the E-2C and S-3B during the 1980s illustrate the Navy's transition toward digitally integrated, network-enabled warfare. Through expanded computing capacity, advanced radar and passive detection systems, and the integration of precision-guided standoff weapons, carrier aviation evolved into a cohesive, technologically sophisticated force structured to confront the high-threat environment of the late Cold War.

This SH-3H Sea King (148039) of HS-9 "Sea Griffins" is seen tied down on the deck of USS Abraham Lincoln (CVN-72) on November 10, 1989, prior to the aircraft carrier's maiden deployment. The aircraft displays newly applied low-visibility markings, which became standard toward the end of the 1980s. (PHOTO: OFFICIAL US NAVY PHOTO BY PH2 TRACY LEE DIDAS)

The Carrier Onboard Delivery (COD)

Carrier Onboard Delivery (COD) aircraft were a vital part of U.S. Navy carrier strike group operations during the 1980s. While frontline aircraft such as the F-14 Tomcat and A-6 Intruder projected combat power, COD aircraft ensured those forces could remain effective by transporting personnel, mail, and high-priority spare parts directly to carriers operating far from shore. During the Cold War and the Reagan administration's "600-ship Navy" expansion, this rapid logistics capability became essential to sustaining global naval deployments.

Three aircraft defined the COD mission in the 1980s: the Grumman C-1A Trader, the Grumman C-2A Greyhound, and the Lockheed US-3A Viking. The C-1A Trader, derived from the S-2 Tracker, represented the final generation of piston-engine COD aircraft. It could carry about nine passengers or 3,500 pounds of cargo, but its reliance on aviation gasoline made it increasingly impractical as the fleet standardised on jet fuel. As a result, the Trader was phased out by 1988.

The primary COD aircraft of the decade was the turboprop-powered C-2A Greyhound. A major reprocurement program beginning in 1984 introduced improved engines, strengthened airframes, and upgraded avionics. The Greyhound could carry up to 26 passengers or 10,000 pounds of cargo, including jet engines for frontline aircraft. Its rear loading ramp and compatibility with carrier deck operations made it indispensable, and by the late 1980s it formed the backbone of fleet logistics support.

Complementing the Greyhound was the jet-powered US-3A Viking, a converted antisubmarine aircraft used mainly in the Western Pacific. Although it carried less cargo, it offered greater speed, longer

EA-3B Skywarrior (BuNo unknown) of VQ-1 "World Watchers" from the January 12, 1986 to August 13, 1986 deployment aboard USS Enterprise (CVN-65) which took CVW-11 around the world. By the 1980s, the Whale was in its sunset years as a carrier-borne platform, nearing the end of a remarkable career that had started in the 1950s. Having served the U.S. Navy for over thirty years in a variety of critical roles, the aircraft was increasingly challenged by advancing technology and the arrival of more modern successors.

range, and aerial refuelling capability. This made it ideal for urgent deliveries across large ocean distances.

The SH-3 Sea King: The multirole workhorse

During the 1980s, the United States Navy relied heavily on the SH-3 Sea King as its primary "inner-zone" anti-submarine warfare platform within carrier battle groups. This decade represented both the peak of the aircraft's operational capability and the beginning of its gradual transition out of frontline combat service. Although it would eventually be replaced by the newer SH-60F Seahawk, the Sea King remained indispensable throughout the 1980s due to a combination of technological upgrades, evolving mission flexibility, and strategic necessity during the late Cold War.

The Sea King's continued relevance was not simply a matter of inertia. Instead, it reflected a deliberate effort by the Navy to modernise an already proven platform. Through incremental upgrades, particularly in digital sensor systems and avionics, the SH-3 evolved into a sophisticated and versatile aircraft capable of meeting increasingly complex operational demands. As a result, the 1980s can be seen as the period when the Sea King reached its highest level of refinement.

At the centre of this evolution was the SH-3H variant, which became the definitive version of the Sea King during the decade. Unlike earlier models that were more narrowly focused on anti-submarine warfare, the SH-3H was designed as a multi-role platform capable of performing several missions simultaneously. These included ASW operations, plane guard duties involving the recovery of downed pilots near aircraft carriers, and general utility or transport tasks.

Most SH-3H helicopters were not newly built but rather converted from earlier variants such as the SH-3G. These conversions began in the 1970s and continued into the 1980s, incorporating increasingly advanced mission systems as technology progressed. This approach allowed the Navy to extend the usefulness of existing airframes while keeping pace with evolving threats.

The SH-3H's flexibility made it particularly valuable in carrier operations, where space, resources, and response times were all critical factors. A single aircraft capable of performing multiple roles reduced logistical strain and increased operational efficiency, making the Sea King an essential component of naval aviation during this period.

Sensor Integration and Tactical Capability

One of the defining features of the SH-3H in the 1980s was its role as a comprehensive sensor platform. By the middle of the decade, the helicopter was equipped with an impressive array of systems that allowed it to detect, track, and classify both surface and subsurface threats.

At the core of its ASW capability was the AN/AQS-138 dipping sonar, which could be lowered into the water to detect submarines. This system was complemented by a set of sonobuoy dispensers located in the rear cabin. These buoys could be deployed over a wide area to create an acoustic detection field, with data transmitted back to the aircraft and onward to the carrier group.

In addition to its underwater detection capabilities, the SH-3H also possessed a surface search radar housed in a retractable radome beneath the fuselage. This radar provided the helicopter with a secondary airborne early warning function, enabling it to identify low-flying aircraft or incoming threats that might otherwise evade detection by ship-based systems.

Electronic Support Measures further enhanced the aircraft's situational awareness by allowing it to detect and analyse enemy radar and radio emissions. This passive capability was particularly valuable in contested environments, where minimising one's own emissions could be critical for survival.

Another important feature was the Magnetic Anomaly Detector, or MAD system, which was used to identify disturbances in the Earth's magnetic field caused by submerged submarines. This system, deployed from the helicopter's sponson, provided a final confirmation tool once a potential target had been localized through other means.

An EA-3B Skywarrior (BuNo unknown) of VQ-1 Det. A "World Watchers" is shown on the catapult aboard the USS Kitty Hawk (CV-63) during the January 3 to June 29, 1987 deployment. A key identifying feature of the EA-3B was the large ventral fairing beneath the fuselage, which housed its ELINT equipment. (PHOTO: OFFICIAL US NAVY PHOTO)

A SH-3H Sea King of HS-4 "Black Knights" is shown hovering over the Pacific Ocean during the 1989 deployment aboard the USS Carl Vinson (CVN-70).

A C-2A Greyhound (BuNo unknown) of VR-24 "Lifting Eagles" is shown landing aboard the USS Coral Sea (CV-43) on March 18, 1986. Visible behind it are aircraft from CVW-13, some of which would likely benefit from the cargo delivered by the Greyhound—whether spare parts or personnel. Just as importantly, the aircraft also carried mail for the ship's crew. In an era before digital communication, carrier onboard delivery (COD) flights were essential not only for sustaining air wing operations but also for maintaining morale among sailors deployed at sea.

Together, these systems transformed the SH-3H into a highly capable and integrated sensor node, able to contribute significantly to the overall tactical picture of a naval task force.

The Shift to Digital: Sonar Advancements

A major technological shift during the early 1980s was the transition from analogue to digital processing in acoustic systems. This change was embodied in the introduction of the AN/AQS-13E dipping sonar, which brought a new level of sophistication to submarine detection.

The AQS-13E incorporated a Sonar Data Computer that allowed for digital processing of both sonar and sonobuoy signals. This advancement significantly improved the accuracy and reliability of target detection, particularly in challenging environments such as shallow coastal waters. One of the key innovations was the use of adaptive signal processing techniques, including Fast Fourier Transform methods, to filter out background noise and reverberation.

These improvements were especially important during the Cold War, when naval operations increasingly took place in littoral zones where acoustic conditions were complex and unpredictable. The ability to distinguish genuine submarine contacts from environmental noise gave the Sea King a critical edge.

Development continued with the AQS-13F, which was intended for the next generation of naval helicopters but was also tested extensively on the Sea King. While trials revealed some technical challenges, such as performance issues at greater depths, the SH-3 platform played an essential role as a testbed for these emerging technologies.

Extending the Life of a Proven Platform

By the late 1980s, it was clear that the Sea King would eventually be replaced, but the transition would take time. To ensure that the fleet remained operationally effective, Sikorsky and the Navy implemented a Service-Life Extension Program beginning in 1987.

This program focused on improving both the durability and survivability of existing aircraft. Mechanical upgrades included modifications to the engines and rotor systems, as well as the addition of features designed to improve performance in adverse conditions, such as ice screens for engine intakes.

Safety enhancements were also a priority. The introduction of crashworthy seating, better soundproofing, and improved emergency lighting made the aircraft safer and more comfortable for its crew. These changes reflected a growing emphasis on human factors in military aviation.

Avionics were updated as well, with some aircraft receiving improved navigation systems and refined sonar processing capabilities. In certain cases, older radar and electronic systems were removed to make room for these newer technologies, illustrating the ongoing process of adaptation and optimization.

A Legacy of Adaptation

The 1980s marked the final chapter of the Sea King's service as a primary combat platform in the United States Navy. However, rather than fading into obsolescence, the aircraft demonstrated a remarkable ability to adapt to changing technological and strategic requirements.

Through digital upgrades, sensor integration, and structural improvements, the Navy was able to extend the operational life of the SH-3 and maintain its effectiveness during a critical period of the Cold War. At the same time, the gradual transition to newer platforms ensured that the Sea King could continue to serve in supporting roles for years to come.

The last Skywarriors

Having entered service in the late 1950s, the A-3 Skywarrior was already approaching the end of its frontline service life by the beginning of the 1980s. The Skywarrior's operational history is a remarkable example of the adaptability of an airframe originally designed as a nuclear bomber, which was later converted to serve in a wide variety of roles, including tanker, VIP transport, electronic warfare, aggressor training, and photo-reconnaissance missions.

The longest-serving variant, however, was the EA-3B Skywarrior, a reconnaissance version equipped with sensors for ELINT operations. While most other variants were gradually retired, the EA-3B remained in service until the early 1990s, when it was finally replaced by the ES-3A Shadow.

The EA-3B carried a crew of seven and was the heaviest aircraft to operate from a U.S. Navy aircraft carrier. Not only was it heavy, but it also occupied considerable space on both the flight deck and the hangar deck. Despite these limitations, the EA-3B provided a unique and valuable capability, which made it common for carrier air wings to include a small detachment of these aircraft as part of their deployment.

Despite its advancing age, the EA-3B played an essential role in both Operation Prairie Fire and Operation El Dorado Canyon. During these operations, it mapped Libyan air defence networks and supported the jamming missions of Grumman EA-6B Prowler aircraft by providing detailed intelligence on radar frequency bands and emitter characteristics.

USS *Nimitz* (CVN-68) returning to Norfolk following the tragic crash of an EA-6B Prowler on the flight deck. The accident claimed the lives of 14 crew members and injured 45 others. It serves as a grim reminder of the hazardous working environment faced by personnel on an aircraft carrier flight deck. (PHOTO: OFFICIAL US NAVY PHOTO BY PH1 KEN BREWER, USN)

SIGNIFICANT EVENTS

The Cold War Carrier Navy

The 1980s heralded a paradigm shift in American naval power, transitioning from a period of post-Vietnam retrenchment to an era of assertive global presence and strategic revitalisation. This decade, often characterised by the overarching framework of the Reagan Doctrine, saw the United States Navy emerge as the primary instrument of national policy, designed to confront the Soviet Union across the maritime domain. Central to this resurgence was the conceptual and physical realisation of the "600-ship Navy," a goal that sought not only to restore numerical parity with the Soviet fleet but to achieve a qualitative margin of superiority that would deter aggression through the credible threat of offensive action. The aircraft carrier battle group served as the centrepiece of this strategy, acting as a mobile, sovereign airfield capable of projecting power into the most contested waters on the globe.

The Reagan Doctrine

The ideological foundations of the 1980s Navy were rooted in a rejection of the "swing strategy" that had dominated naval planning during the late 1970s. Under the Carter administration, naval strategy was largely reactive, focusing on the defence of the Atlantic Sea Lines of Communication (SLOC) to ensure the reinforcement of NATO forces in Europe in the event of a Warsaw Pact invasion. This defensive posture effectively ceded the initiative

An F/A-18A Hornet (BuNo unknown) of VFA-113 "Stingers" is seen seconds before being catapulted from the deck of USS Constellation (CV-64) in September 1989 during Exercise PACEX '89. The deployment lasted from September 16 to October 19, 1989, and took CVW-14 to the Norpac. (PHOTO: OFFICIAL US NAVY PHOTO)

to the Soviet Navy, particularly in the Pacific and the high north, where Soviet "bastions" protected their ballistic missile submarines from Western interference.

The election of Ronald Reagan in 1980 brought a new strategic imperative. The Reagan Doctrine aimed to roll back communist influence globally by providing support to anti-communist insurgencies and modernising the U.S. military to a level that would make the cost of Soviet competition unsustainable. National Security Decision Directive 32 (NSDD-32), issued in March 1982, formalised this intent by mandating a global, forward-leaning military posture. For the Navy, this meant transitioning from a role of "support" to one of "command of the seas," a shift championed by Secretary of the Navy John Lehman and Chief of Naval Operations (CNO) Admiral James Watkins.

The Mediterranean Powder Keg

Throughout the 1980s, the Mediterranean Sea was the primary stage for a high-stakes chess match between *American* naval power and regional defiance. At the centre of this storm was Muammar Gaddafi's Libya, a regime that challenged international norms through state-sponsored terrorism and bold territorial claims over the Gulf of Sidra.

By drawing a "Line of Death" across international waters, Gaddafi set the stage for a decade of kinetic confrontation that would push the U.S. Navy's Cold War doctrine to its limit.

A EA-3B Skywarrior (146454) from VQ-2 "Batmen" launches from USS *Dwight D. Eisenhower* (CVN-69) during the April 27, 1983 to December 2, 1983 deployment to the Mediterranean Sea. (PHOTO: OFFICIAL US NAVY PHOTO)

An F/A-18A Hornet (162901) of VFA-192 "Golden Dragons" in flight over USS *Midway* (CV-41) in June 1989. (PHOTO: OFFICIAL US NAVY PHOTO BY LCDR T.B. SURBRIDGE, USN)

n E-2C Hawkeye (BuNo unknown) of VAW-123 "Screwtops" is seen awaiting transfer to the flight deck via one of the aircraft elevators aboard USS *heodore Roosevelt* (CVN-71). The "Screwtops" were assigned to CVW-1, which operated from USS America (CV-66). However, during USS Theodore oosevelt's initial shakedown cruise, squadrons from three different air wings were embarked, the majority of them from CVW-1.
HOTO: OFFICIAL US NAVY PHOTO BY PHAN SARNA)

Testing the "Line of Death"

The decade was defined by a series of escalating "Flashpoints" that transformed the Mediterranean into a live-fire laboratory for modern naval warfare.

From the first post-Vietnam air-to-air kills in 1981 to the high-tech intercepts of 1989, Navy F-14 Tomcats consistently demonstrated their dominance in aerial dogfights against Libyan Fitters and Floggers, reinforcing their reputation as formidable air superiority fighters. At the same time, large-scale operations such as Prairie Fire and El Dorado Canyon highlighted the devastating effectiveness of integrated carrier warfare, where multi-carrier battle groups combined Aegis cruisers, electronic warfare capabilities, and precision strike assets to systematically dismantle Libyan defences. Yet, these successes were tempered by the harsh realities of command and control, as not every mission unfolded as planned; the 1983 strikes over Lebanon, in particular, underscored the risks of rigid bureaucracy and tactical interference, resulting in downed aircraft and a highly visible diplomatic crisis in the Bekaa Valley.

These engagements were more than just isolated skirmishes; they were the crucible in which modern naval aviation was refined.

The hangar deck of USS *Constellation* (CV-64) is seen on October 29, 1985. Visible in the photograph are an F/A-18A Hornet from VFA-25 "Fists of the Fleet," S-3A Vikings from VS-37 "Sawbucks," and the tail section of an E-2C Hawkeye from VAW-113 "Black Eagles."
(PHOTO: OFFICIAL US NAVY PHOTO BY JO2 MILLIE TAMBERG)

An F-14A Tomcat of VF-24 "Fighting Renegades" flies alongside F-4S Phantoms of VF-301 "Devil's Disciples" near Yuma in March 1986. The image is almost symbolic, depicting the Tomcat accompanied by two of its predecessors. It evokes the scene of a dissimilar air combat maneuvering (ACM) exercise in which the newer F-14 may well have demonstrated the advantages that made it the Navy's premier fleet defense fighter of its era. (PHOTO: OFFICIAL US NAVY PHOTO BY LCDR DAVID BARANAK)

An EA-6B Prowler of VAQ-139 "Cougars" is seen seconds before launch from USS Constellation (CV-64) on January 15, 1987, during carrier qualifications (CarQuals) for an upcoming deployment. That deployment, which lasted from April 11 to October 13, 1987, took USS *Constellation* and CVW-14 to the Westpac., the Indian Ocean, and the Persian Gulf. (PHOTO: OFFICIAL US NAVY PHOTO BY JOC RICH BETH, USNR-R)

THE GULF OF SIDRA SHOWDOWNS

The Black Aces and the Sukhoi Engagement

In mid-August 1981, the U.S. Sixth Fleet conducted an Open Ocean Missile Exercise (OOMEX) in the Mediterranean. A force consisting of USS *Nimitz* (CVN-68) and USS *Forrestal* (CV-59) was commanded to establish a clear line of fire for live missile tests while simultaneously challenging Gaddafi's territorial assertions, near the so-called "Line of death".

On the first day of Exercise , August 18, the Libyan Arab Air Force (LAAF) scrambled no fewer than 35 pairs of interceptors, including MiG-23 "Floggers," MiG-25 "*Fox*bats," and Mirage F1s. These aircraft attempted to penetrate the U.S. exercise zone in order to locate the carriers.

U.S. Navy F-14A Tomcats from VF-41 "Black Aces" and VF-84 "Jolly Rogers," operating from USS *Nimitz* (CVN-68), along with F-4J Phantoms from VF-74 "Be-Devilers" and VF-103 "Sluggers," operating from USS *Forrestal* (CV-59), were tasked with establishing a Barrier Combat Air Patrol (BARCAP) to intercept these intruders.

During these initial encounters, the tactical superiority of the F-14 was evident. One specific engagement involved VF-41 Tomcats intercepting a pair of MiG-25s. The *American* pilots managed to evade the MiGs' radar with the aid of radar jamming by nearby EA-6B and EP-3 and the MiGs' limited situational awareness. The Tomcats positioned themselves directly behind the *Fox*bats and then accelerated to fly alongside them. In a moment of psychological warfare that later achieved cultural immortality, the Radar Intercept Officer (RIO) of the wingman Tomcat, "Fast Eagle 107," issued an informal "rude hand gesture" to the unsuspecting Libyan pilot.

The S-3A Probe

The following morning, August 19, 1981, the Navy executed a tactical probe to evaluate Libyan reaction times. An S-3A Viking from VS-24 "Scouts" operating from USS *Nimitz* (CVN-68) was dispatched to fly off the coast, acting as a "honey pot" for Libyan radar. While the Viking flew its pattern, an E-2C Hawkeye from VAW-124 "Bear Aces", also operating from USS *Nimitz* (CVN-68) maintained overwatch from a high-altitude station, its APS-125 radar scanning deep into Libyan territory.

At approximately 07:00 local time, the E-2C detected two Sukhoi Su-22

F-4S Phantom II (155825) of VF-103 "Sluggers" from the June 7, 1982 to November 16, 1982 deployment aboard USS Forrestal (CV-59) in the Mediterranean Sea and Indian Ocean. The final version of the Phantom II to serve from U.S. aircraft carriers was the F-4S, an upgraded F-4J. The Sluggers operated this variant only on this deployment and transitioned to F-14 Tomcats soon after returning to NAS Oceana in late 1982.

An F-14A Tomcat (BuNo unknown) of VF-41 "Black Aces" immediately after landing aboard USS Nimitz (CVN-68) in August 1981. This photograph was taken around the time of the first Gulf of Sidra engagement; however, this particular aircraft was not directly involved in the incident.
(PHOTO: OFFICIAL US NAVY PHOTO BY PH3 CRUZ)

Two F-14A Tomcat aircraft (BuNos 160403 and 160390) from VF-41 "Black Aces" are depicted as they appeared during the events of August 19, 1981. Aircraft 160403 is shown in its pre-mission configuration, while 160390 is represented in its post-mission appearance. Note the Sukhoi Su-22 kill marking added beneath the tail flash following the engagement with Libyan aircraft.

"Fitter-J" fighter-bombers launching from Ghordabiya Air Base. Due to their tight formation, the Libyan aircraft initially appeared on the Hawkeye's radar as a single contact. The S-3A, lacking its own threat warning receiver at the time, was immediately vectored away from the intercept path, while two F-14As from VF-41 were dispatched to meet the incoming threats.

Fast Eagle 102 and 107

The Tomcat section was composed of:

1. Fast Eagle 102 (160403): Piloted by Cdr. Henry "Hank" Kleemann (CO) with Lt. David "DJ" Venlet as RIO.

2. Fast Eagle 107 (160390): Piloted by Lt. Lawrence "Music" Muczynski with Lt.(jg) James "Luca" Anderson as RIO.

As the Tomcats closed to within 60 miles of the Libyan coast, they obtained a radar lock on the bogeys. At eight miles, the crews achieved visual identification of the two Su-22s flying in a "welded wing" formation, separated by only about 150 feet. While previous intercepts had typically ended with the Libyan aircraft turning back, these Fitters continued their head-on approach.

As the aircraft crossed at a distance of roughly 1,000 feet, the lead Libyan pilot fired an AA-2 "Atoll" (K-13) infra-red seeking missile at Fast Eagle 102. Kleemann observed the missile launch from the Su-22's right wing station and immediately communicated the hostile act to his wingman. The Atoll missile, poorly suited for a head-on engagement against a target with the Tomcat's heat signature and manoeuvrability, failed to track and passed harmlessly.

The Response

Under the prevailing Rules of Engagement (ROE), the initiation of hostile fire by the Libyans granted the Tomcat crews the authority to defend themselves. The Su-22s executed a standard "shell" defensive split, with the leader turning northwest and the wingman turning southeast toward the coast.

The Tomcats, benefitting from the F-14A's superior manoeuvrability easily out-turned the Libyans. Cdr. Kleemann pursued the Su-22 leader, manoeuvring into the "six o'clock" position and launching an AIM-9L Sidewinder. The missile struck the Sukhoi's tailpipe, resulting in a catastrophic explosion. Simultaneously, Lt. Muczynski in Fast Eagle 107 pursued the wingman. After receiving confirmation from his leader to engage, Muczynski launched his own AIM-9L Sidewinder, which also scored a direct hit.

Both Libyan pilots successfully ejected from their aircraft and were seen with "good chutes" by the *American* crews. The entire engagement, from the Libyan missile launch to the second shootdown, lasted approximately 45 seconds. The radio transmission "102, 107. Two enemy kills" signalled the first combat victory for the F-14 and the first air-to-air victory for the U.S. Navy since 1973.

The Air Battle Near Tobruk

The final major air-to-air confrontation of the Cold War era occurred on January 4, 1989. Tensions had spiked due to the construction of a chemical weapons facility at Rabta, which the U.S. believed was intended for the production of nerve gas. The USS *John F. Kennedy* (CV-67) was operating north of the Libyan coast when a pair of MiG-23 "Floggers" launched from

The F-14A Tomcat (159610) of VF-32 "Swordsmen," flown by Cdr. Joseph Bernard Connelly and Cdr. Leo F. Enwright, is depicted as it appeared on January 4, 1989, following the mission in which they shot down a MiG-23. Note the MiG-23 silhouette applied just above the modex number to commemorate the aerial victory. As seen in photographs taken during the homecoming, this kill marking had been removed before the deployment concluded.

Al Bumbah airfield and headed directly toward the battle group.

Gypsy 207 and Gypsy 202

The intercepting section came from VF-32 "Swordsmen":

1. Gypsy 207 (159610): Crewed by Cdr. Joseph Bernard Connelly (CO) and Cdr. Leo F. Enwright (RIO).

2. Gypsy 202 (159437): Crewed by Lt. Hermon C. Cook III (Pilot) and Lt.Cdr Steven Patrick Collins (RIO).

An E-2C Hawkeye, call sign "Closeout," detected the MiG-23s at a distance of 72 nautical miles. Unlike the 1981 incident, which was a close-quarters ambush, the 1989 engagement was a prolonged tactical chess match that highlighted the evolution of U.S. Navy Air Combat Manoeuvring (ACM) and Rules of Engagement.

Tactical Manoeuvres

As the MiG-23s closed, the Tomcat crews attempted to resolve the situation without violence, in accordance with the "peacetime" ROE which emphasised avoiding the first use of force. The Tomcats initiated a 30-degree offset turn away from the MiGs to signal that they were not on an attack profile. However, the Libyan aircraft repeatedly "jinked" back toward the Tomcats to maintain a head-on, "nose-on" intercept.

To negate the threat of the MiG-23's AA-7 "Apex" missiles—which had a potential range of 12 miles—the F-14s descended from 20,000 feet to 3,000 feet. This altitude drop served two purposes: it allowed the F-14s to use the ocean surface as radar "clutter" to confuse the MiGs' simpler radar systems and exploited the "Doppler Notch". By "beaming" (flying perpendicular to) the MiGs' radar beam, the Tomcats could become invisible to the

"Gypsy 207" (159610) is seen returning to NAS Oceana on January 31, 1989, following deployment aboard USS John F. Kennedy (CV-67). This MiG-23 "killer" was flown by Cdr. Joseph Bernard Connelly and Cdr. Leo F. Enwright during the notable mission of January 4, 1989. (PHOTO: OFFICIAL US NAVY PHOTO BY PH1 JEFF ELLIOT)

An E-2C Hawkeye (162616) of VAW- 126 "Seahawks," deployed aboard the USS John F. Kennedy (CV-67) from August 2, 1988 to February 1, 1989 in the Mediterranean Sea. This aircraft played an important role in engagements such as the one on January 4, 1989, by enhancing the situational awareness of F-14 Tomcat crews through airborne early warning and battle management support.

MiG's weapons-tracking software, which was designed to filter out objects with low relative radial velocity. The "Dopper Notch" reduces the target's relative velocity to zero, causing the radar to filter the aircraft out as "ground clutter," losing lock.

The Engagement Sequence

After the MiGs turned back toward the Tomcats for a fifth time, the air warfare commander aboard the USS *John F. Kennedy* granted authority to fire in self-defence. The ROE had evolved since 1981; the repeated aggressive manoeuvring by the Libyans was now classified as "hostile intent," allowing the *Americans* to strike first to ensure their own survival.

The engagement proceeded as follows:

At ranges of 12 miles and again at 10 miles, Enwright in Gypsy 207 initiated the engagement by firing AIM-7 Sparrow missiles, but both failed to track their targets—likely due to technical issues or the challenging geometry of the intercept. As the distance closed to five miles, the Tomcats executed a defensive split, breaking in opposite directions to force the MiGs to commit to a single target; both enemy aircraft chose to pursue Gypsy 202. Seizing the opportunity, Cook launched an AIM-7 Sparrow from roughly five miles, achieving a direct hit that destroyed the first MiG-23. Meanwhile, Cdr. Connelly in Gypsy 207 maneuverer into the rear quadrant of the remaining MiG, working to secure a heat-seeking lock; after briefly struggling to acquire a proper tone, he fired an AIM-9L Sidewinder from about one mile, and the missile struck decisively, detonating in the MiG's tailpipe and eliminating the second aircraft.

Both Libyan pilots successfully ejected, and the VF-32 crews confirmed "two good chutes" before returning to the carrier. The engagement lasted eight minutes from initial detection to the final kill.

F-14A Tomcat (161283) from VF-102 "Diamondbacks" from the March 10, 1986 to September 10, 1986 deployment aboard USS America (CV-66) in the Mediterranean Sea. This Tomcat is shown here in the 4×4 missile configuration, carrying four AIM-9 Sidewinder and four AIM-7 Sparrow missiles. This was a typical loadout for the squadron, whereas its sister squadron, VF-33 "Starfighters," more commonly employed the 2×2×2 configuration, consisting of two AIM-9 Sidewinders, two AIM-7 Sparrows, and two AIM-54 Phoenix missiles.

OPERATION EL DORADO CANYON

The maritime theatre of the Mediterranean in the mid-1980s served as the primary arena for a transformative confrontation between the United States and Libya. While Operation El Dorado Canyon is frequently characterised in historical memory by the long-distance sorties of United States Air Force F-111 Aardvarks, the structural integrity and tactical success of the mission were fundamentally underpinned by the United States Navy's carrier-based assets. Operating from the aircraft carriers of the Sixth Fleet, US Naval aviation executed a multi-phased campaign that integrated advanced electronic warfare, precision strike, and air superiority to challenge Libyan territorial overreach and dismantle the infrastructure of state-sponsored terrorism.

Background

The escalation of 1986 was not an isolated phenomenon but the culmination of two decades of deteriorating relations. Since the 1969 coup that brought Muammar Qaddafi to power, Libya had increasingly positioned itself as a revolutionary vanguard, often utilising terrorism as an instrument of state policy. By the early 1980s, the Reagan administration had shifted its counter-terrorism strategy from a law enforcement paradigm to one of military deterrence.

Central to the maritime friction was Qaddafi's unilateral assertion of the "Line of Death," a boundary established at 32°30′ North latitude. By declaring the entire Gulf of Sidra as Libyan internal waters—a claim that defied the 12-nautical-mile limit codified in international law—Libya sought to exclude the United States from 3,200 square miles of international waters. The US Navy's response, manifested through Freedom of Navigation (FON) operations and the "Attain Document" series of exercises, was designed to physically challenge this boundary and assert the right of international passage.

Operation Prairie Fire

In March 1986, the simmering tensions erupted into open conflict during Operation Prairie Fire. This engagement served as a vital proving ground for the coordinated application of the Carrier Air Wing (CVW) against a sophisticated, Soviet-equipped adversary. The involvement of three carriers—USS *Saratoga* (CV-60), USS *America* (CV-66), and USS *Coral Sea* (CV-43)—provided the Sixth Fleet with a concentrated force of approximately 250 aircraft.

Surface-to-Air Missile Confrontations

The threshold of kinetic engagement was crossed on March 24, 1986, at 07:52 CET. A Libyan missile installation near Sirte launched two Soviet-made SA-5 "Gammon" (S-200) missiles at F-14A Tomcats from the USS *America*'s VF-102. These missiles, designed for high-altitude, long-range interception, failed to acquire their targets and fell into the Mediterranean. A second salvo was

A CH-46 Sea Knight delivers supplies to USS Saratoga (CV-60) on March 20, 1986. Visible in the photograph are an F-14A Tomcat from VF-74 "Be-Devilers," an A-7E Corsair II from either VA-81 "Sunliners" or VA-83 "Rampagers," an S-3A Viking from VS-30 "Diamondcutters," and an A-6E along with two KA-6D Intruders from VA-85 "Black Falcons." USS Saratoga, with CVW-17 embarked, was deployed to the Mediterranean Sea from August 26, 1985, to April 16, 1986, and flew missions in support of Operation Prairie Fire.
PHOTO: OFFICIAL US NAVY PHOTO BY PH1 WILLIAM H. DODGE

EA-6B Prowler (159908) from VMAQ-2 Det. Y "Playboys" from the March 10, 1986 to September 10, 1986 deployment aboard USS America (CV-66) in the Mediterranean Sea. Equipped with three AN/ALQ-99 jamming pods, this EA-6B Prowler provided extensive electronic warfare support before, during, and after Operation El Dorado Canyon, significantly enhancing the protection of American aircraft against Libya's integrated air defence system.

aunched later that afternoon against Tomcats on the southernmost combat air patrol (CAP). In this instance, the electronic warfare capabilities of the EA-6B Prowler of VMAQ-2 Detachment Y were decisively demonstrated; the Prowler's jamming suites effectively neutralised the Libyan tracking radars, causing the missiles to ditch.

The Whale

Libya's air defence network in 1986 was a formidable array of Soviet-supplied technology, overseen by Soviet and East German advisers. This network was concentrated around the major population centres and military hubs of Tripoli and Benghazi. The challenge for U.S. planners was to penetrate the Libyan IADS—one of the densest in the world at the time—without incurring prohibitive losses.

The presence of the SA-5 Gammon was particularly concerning for U.S. Navy and Air Force planners. With its massive range, the SA-5 could target the high-altitude tankers and command aircraft that were essential for the long-range F-111 strike from the United Kingdom. Mapping

Two F-14A Tomcats of VF-33 "Starfighters" are seen on the deck of USS America (CV-66) on April 17, 1986, just days after Operation El Dorado Canyon. Although the operation had concluded, the flight deck remained highly active, with patrol missions being conducted at a rapid pace in anticipation of potential retaliatory actions by remaining Libyan forces. (PHOTO: OFFICIAL US NAVY PHOTO BY PHAN DAVID CASPER)

the precise locations and operating frequencies of these systems fell largely to the EA-3B Skywarriors of VQ-2 operating off the deck of USS *America* (CV-66). The "Whale" was the only platform that could loiter at sufficient altitude and distance to safely collect the sensitive electronic signatures required to build a comprehensive Electronic Order of Battle.

In the weeks preceding the April 15 raid, VQ-2 Skywarriors were deeply involved in Operation Prairie Fire and subsequent manoeuvres in the Gulf of Sidra. These operations were designed to be provocative; by sending carrier-based fighters toward the "Line of Death," the U.S. forced Libyan air defence operators to activate their tracking and guidance radars. When a Libyan radar "painted" a U.S. aircraft, the EA-3B was there to record the signal, analyse its pulse repetition frequency (PRF), pulse width, and frequency agility.

This data was transmitted back to the National Security Agency and Naval Intelligence for detailed analysis. The resulting intelligence was used to create "threat libraries" for the ALQ-99 jamming pods on the EA-6B Prowlers and EF-111 Ravens. Without the precise technical data captured by the EA-3B, the active jamming efforts would have been significantly less effective, as they would have had to rely on generalised "barrage" jamming rather than the "spot" or "deceptive" jamming required to blind specific Soviet-made systems.

An EA-3B Skywarrior (146555) of VQ-2 "Batmen" is seen preparing to land aboard USS Coral Sea (CV-43) on March 18, 1986. Two EA-3Bs operated under the command of CVW-13 for part of the deployment, which lasted from October 1, 1985, to May 18, 1986. The role of the Skywarriors during Operation El Dorado Canyon cannot be overstated, as they played a critical part in ensuring the safety of the strike package. (PHOTO: OFFICIAL US NAVY PHOTO BY PH2 R.R. KNEPP)

Even in the twilight of its career, "The Whale" continued to play a crucial role in carrier air wings by providing valuable intelligence for mission planning and post-mission evaluation. Although originally designed as a nuclear bomber, the aircraft went on to serve in a wide range of additional roles. In its reconnaissance capacity in particular, it remained indispensable and was not replaced until the early 1990s.

The whale continued to play a crucial role during Operation El Dorado Canyon as well as in the aftermath of the mission.

Aerial Supremacy and the MiG-25 Intercept

The disparity in situational awareness and pilot proficiency was most evident in the air-to-air arena. When two Libyan MiG-25 *Foxbats* launched from Benina Air Base with orders to intercept US fighters, they were immediately detected by E-2C Hawkeye early warning aircraft. The Hawkeyes vectored F-14A Tomcats from VF-33 to intercept. During the ensuing engagement at 5,000 feet, the Tomcat pilots utilised their superior low-altitude manoeuvrability to achieve tactical dominance over the high-speed, but less agile, MiGs. By positioning themselves behind the MiGs and acquiring "Sidewinder tones," the Navy pilots demonstrated the ability to destroy the Libyan aircraft at will. Although no missiles were fired due to the specific rules of engagement at the time, the retreat of the MiGs signalled the Navy's absolute control of the airspace over the Gulf of Sidra.

The Debut of the Harpoon

Operation Prairie Fire also marked the combat debut of the AGM-84 Harpoon anti-ship missile. When Libyan patrol boats moved aggressively toward the carrier task force, the Navy responded with a combination of surface and aerial strikes. On March 24th and 25th A-6E Intruders from VA-34, VA-55 and VA-85 fired Harpoon missiles and damaged or sank one Libyan patrol boat and two Libyan Corvettes.

These actions were supported by A-7E Corsair IIs from VA-83, which employed the AGM-88 HARM (High-speed Anti-Radiation Missile) to damage Libyan radar sites at Sirte. The successful suppression of these long-range sensors was a prerequisite for the more ambitious strikes planned for the following month.

The Raid on Benghazi and Benina

In the early hours of April 15, 1986, the Sixth Fleet launched a highly synchronised strike package coordinated to hit targets in eastern Libya simultaneously with the Air Force's strikes on Tripoli. The operation utilised a complex array of deconfliction procedures to ensure that Navy aircraft arriving from the Mediterranean were over their targets at the same second as the Air Force bombers arriving from the United Kingdom.

Benina and Benghazi air strikes

The strike on the Benina Airfield, located on the outskirts of Benghazi, was a critical objective intended to neutralise Libya's primary fighter base in the eastern sector.

A-7E Corsair II (159660) of VA-46 "Clansmen" from the March 10, 1986 to September 10, 1986 deployment aboard USS America (CV-66) in the Mediterranean Sea. Armed with the AGM-88 HARM anti-radiation missile, this aircraft participated in the suppression of Libyan air defences during the missions leading up to Operation El Dorado Canyon.

A-7E Corsair II (159283) of VA-83 "Rampagers" from the August 25, 1985 to April 16, 1986 deployment aboard USS Saratoga (CV-60) in the Mediterranean Sea. This aircraft participated in the March 24 strikes against Libyan radar sites. The targets were destroyed using the sophisticated AGM-88 HARM anti-radiation missile, one of which can be seen mounted on the aircraft's No. 2 station.

A-6E Intruder (158538) of VA-34 "Blue Blasters" from the March 10, 1986 to September 10, 1986 deployment aboard USS America (CV-66) in the Mediterranean Sea. Armed with CBU-59 APAM cluster bombs and AGM-84 Harpoon anti-ship missiles, this A-6 participated in the anti-ship strike missions conducted during Operation Prairie Fire. One month later, the same aircraft from VA-34 also took part in the airstrikes against the airfield at Benghazi.

An A-7E Corsair II of VA-46 "Clansmen" fires an AGM-88 HARM missile at Libyan air defence positions during Operation Prairie Fire. This operation marked the combat debut of the AGM-88 HARM missile.

A-6E and KA-6D Intruders of VA-55 "Warhorses" are being prepared for the Benina Airfield strike on April 15, 1986. The flight deck is crowded with bomb skids carrying CBU-59 cluster bombs, which were being loaded onto the strike package tasked with anti-personnel and anti-material missions.
(PHOTO: OFFICIAL US NAVY PHOTO BY PHAN FARMER, USN)

The VA-55 "Warhorses" from the USS *Coral Sea* (CV-43) provided the core strike force, launching eight A-6E Intruders.

The ordnance configuration for this mission was meticulously tailored to the target set:

- **Anti-Personnel/Anti-Material Strike:** Five A-6Es were each loaded with 14 CBU-59 APAM cluster bombs. These weapons were designed to blanket the airfield's ramp and alert pads with thousands of high-velocity fragments, destroying MiG-23 Floggers in their revetments and neutralizing ground crews.
- **Structural Destruction:** Three A-6Es were configured with 16 Mk 82 Snakeye high-drag bombs. These bombs, equipped with parachute retarders, allowed for extremely low-altitude delivery while giving the aircraft time to escape the blast radius. They were targeted at hangars and warehouses believed to contain aircraft spares and logistics.

The Intruders executed a "Stream Raid" tactic, entering the target area with one-minute separation at altitudes as low

KA-6D Intruder (149940) of VA-34 "Blue Blasters" from the March 10, 1986 to September 10, 1986 deployment aboard USS America (CV-66) in the Mediterranean Sea. Often an unsung hero of carrier operations, the tanker variant of the Intruder played an important role in both routine flight operations and combat missions such as Operation Prairie Fire and Operation El Dorado Canyon. By providing in-flight refuelling support, it extended the range and endurance of strike aircraft and helped recovering aircraft safely return to the carrier when they were forced to wait for an opportunity to trap on a busy flight deck. Interestingly, the tanker variant was often painted in high visibility colours instead of the subdued TPS colours of the all-weather attack variant.

An A-6E Intruder of VA-34 "Blue Blasters" is shown flying at low level over downtown Benghazi during strike operations. Although unconfirmed, one account from the mission describes a surface-to-air missile pursuing the A-6 and striking a residential building in the city. Other reports suggest that the damage may instead have been caused by bombs released by attacking Intruders. What is certain, however, is that details can easily become unclear in the fog of war.

SH-3H Sea King (unknown) of HS-11 "Dragon Slayers" from the March 10, 1986 to September 10, 1986 deployment aboard USS America (CV-66) in the Mediterranean Sea. "Always prepare for the worst." This helicopter stood as a testament to that principle, as it was tasked with recovering downed U.S. aircrew from either the mainland of Libya or its coastal waters in the event that any personnel were forced to eject during the mission. The camouflage scheme was applied using water-based paint and was intended to resemble that of Libyan Super Frelon helicopters, which carried similar colours and patterns. This temporary disguise was meant to reduce the likelihood that Libyan forces would recognise and engage the rescue aircraft while it conducted combat search and rescue operations.

as 200 feet. Despite the airfield being fully illuminated upon arrival, the Libyan response was delayed by intense jamming from the EA-6B and EF-111 packages. The results were catastrophic for the Libyan Air Force: multiple MiG-23s, Mi-8 helicopters, and at least one Fokker F-27 were confirmed destroyed on the ground. A mechanical failure on one aircraft resulted in the jettison of a centreline refuelling pod, which inadvertently struck a warehouse and ignited a massive fire, further damaging the facility.

The second major target in the Benghazi sector was the Jamahiriya Guard barracks, an alternative command and control headquarters. The VA-34 "Blue Blasters," operating from the USS *America* (CV-66), launched six A-6E TRAM Intruders for this high-risk mission into a densely populated urban area.

The strike was conducted using a high-speed, low-level nighttime ingress, a tactic that had become the hallmark of Intruder operations. Four aircraft were equipped with sixteen Mk 82 Snakeye bombs each, while the remaining two aircraft carried eight Mk 82s apiece.

Despite encountering intense anti-aircraft artillery (AAA) fire and numerous surface-to-air missile (SAM) launches, all Blaster crews were engaged—fortunately without any successful enemy hits. Estimates of enemy missile activity ranged from "too many SAMs to count" to "at least several dozen." Blaster Five notably reported that "it looked like the Fourth of July on a grand scale." Several crews also observed multiple SAM impacts on the ground in and around the city.

The Blue Blasters delivered their ordnance with precision, successfully striking both the barracks and aircraft storage facilities. The squadron returned

An F/A-18A Hornet (BuNo unknown) of VFA-132 "Privateers" is seen on the deck of USS Coral Sea (CV-43) on April 15, 1986, shortly before departing on SEAD missions as part of Operation El Dorado Canyon. Note the heavily loaded configuration of the aircraft, carrying both drop tanks, AGM-88A HARM missiles, and AIM-9L Sidewinders. This loadout was too heavy for carrier recovery, indicating that the aircraft was armed with the expectation that at least some of the ordnance would be expended during the mission. (PHOTO: OFFICIAL US NAVY PHOTO BY PHAN FARMER)

FA-18A Hornet (162443) of VFA-132 "Privateers" on the October 1, 1985 to May 19, 1986 deployment aboard USS Coral Sea (CV-43) in the Mediterranean Sea. The four Hornet squadrons of CVW-13, including VFA-132 "Privateers", operating from the deck of the USS Coral Sea (CV-43), conducted suppression of enemy air defenses (SEAD) missions during Operation El Dorado Canyon by firing AGM-88 HARM anti-radiation missiles against Libyan air defence sites.

to the carrier without sustaining damage, demonstrating a high level of tactical expertise and combat readiness.

The success of the A-6E bombing runs was made possible by the comprehensive SEAD coverage provided by F/A-18A Hornets and A-7E Corsairs. Operation Prairie Fire and El Dorado Canyon, marked the first combat employment of the AGM-88 HARM, a weapon that would revolutionize electronic warfare.

As the strike aircraft approached their coast-in points, the SEAD package monitored the electronic spectrum for Libyan radar activity. When the Libyans activated their "Square Pair" and "Bar Lock" radars to guide their SAMs, the Navy and Marine Corps aircraft immediately fired HARMs. Over 48 anti-radiation missiles were fired during the raid, destroying more than a dozen radar systems. This prevented the Libyan air defences from achieving effective locks on the Intruders, forcing them to fire missiles unguided, which significantly reduced the risk to US personnel. The SEAD missions were carried out by Hornets from CVW-13 operating off the deck of USS *Coral Sea* (CV-43) namely VFA-131, VFA-132, VMFA-314 and VMFA-323. From USS *America* (CV-66) A-7E Corsair IIs from VA-46 and VA-72 conducted the SEAD missions, while EA-6B Prowlers of VAQ-135 and VMAQ-2 Det. Y jammed the Libyan air defence radars.

Aftermath

Following the 02:00 strikes on April 15, the mission transitioned to the Battle Damage Assessment (BDA) phase. While SR-71 Blackbirds launched from the UK and provided high-resolution photographic evidence of the destruction at Tripoli and Benghazi, the EA-3Bs of VQ-2 continued

An A-6E Intruder (BuNo unknown) of VA-55 "Warhorses" is seen on final approach to trap aboard USS Coral Sea (CV-43) on March 22, 1986, during Operation Prairie Fire. This image highlights the remarkable versatility of the A-6 Intruder, as the aircraft is armed with an AGM-84 Harpoon missile, two CBU-59 APAM cluster bombs, and, notably, an AIM-9L Sidewinder missile, which was carried for self-defence. (PHOTO: OFFICIAL US NAVY PHOTO BY PHC DUNIVIN)

their electronic surveillance to monitor the status of the Libyan military.

In the hours and days following the raid, the EA-3B's mission was to determine which components of the Libyan IADS were still functional. Many Libyan radar operators, terrified by the effectiveness of the HARM missiles, refused to activate their systems even when U.S. aircraft were present. The EA-3B operators monitored for the sporadic "brief activations" that indicated a radar unit was being tested or moved.

The intelligence gathered by the Skywarriors confirmed that the Libyan communication and defence network was paralysed. The combination of physical destruction at the command centres and the psychological impact of the SEAD effort had effectively neutralised Libya's ability to defend its airspace. This allowed subsequent U.S. Navy patrols to operate with near impunity, further isolating Qaddafi and reinforcing the strategic message of the raid.

In the aftermath of the raid, a significant inter-service debate emerged regarding the necessity of the Air Force's involvement. It was argued that the entire operation could have been conducted using Navy assets alone, given the presence of two carriers on station. Proponents of this view pointed to the inherent flexibility of the aircraft carrier as a "sovereign base" that did not require the permission of third-party nations for overflight or basing—a permission that was notably denied by France, Spain, and Italy during the mission planning phase. However, the joint-service structure was ultimately chosen for its political symbolism and the added firepower provided by the F-111's heavier bomb loads.

EA-6B Prowler (158040) of VAQ-135 "Black Ravens" on the August 22, 1982 to November 4, 1982 deployment aboard USS America (CV-66) to the Norlant. Deploying with the carrier battle group into the North Atlantic barrier region, the squadron contributed electronic warfare coverage against Soviet reconnaissance aircraft and strike assets attempting to locate NATO naval forces transiting the Greenland–Iceland–United Kingdom gap.

NORTH ATLANTIC AND THE GIUK GAP

The Greenland-Iceland-United Kingdom (GIUK) Gap and the Norwegian Sea were in the 1980s emerging as the critical strategic chokepoints of the decade. The GIUK Gap was viewed as a "strategic system" designed to regulate the movement of the Soviet Northern Fleet into the open Atlantic.

The GIUK Gap functioned as a "revolving door" for Soviet submarines transiting from their bases on the Kola Peninsula to their patrol areas in the Atlantic. To counter this, the U.S. Navy and its NATO allies built an entire architecture of surveillance and response around this geography. This included the Sound Surveillance System (SOSUS), which utilised fixed bottom arrays to listen for the distinctive acoustic signatures of Soviet vessels. The western anchor of this gap was Greenland, which provided essential facilities like Pituffik Space Base (formerly Thule) for missile warning and space surveillance. Iceland sat at the centre, hosting P-3 Orion maritime patrol aircraft at Keflavík, while the United Kingdom provided the eastern anchor with its Royal Navy and RAF assets.

For the U.S. Navy, the strategic goal was not merely to monitor the gap but to move north of it. This was a significant turnabout from the 1970s, during which Western navies had largely planned for a "passive defence" behind the GIUK line. In the 1980s, the Maritime Strategy dictated that NATO forces would arrive early at scenes of battle and harness the geography of the High North to win a decisive victory at sea.

The Defence of the Northern Flank

Norway was the focal point of NATO's northern flank throughout the 1980s. Strategists recognised that the loss of northern Norway would be decisive in the battle for the Atlantic, as it would allow the Soviet Union to deploy land-based air cover deep into the Norwegian Sea. Consequently, the support of the land battle in Norway by naval and Marine forces was deemed critical.

The concept of maritime operations (CONMAROPS), adopted by NATO in 1980, highlighted the importance of "containing Soviet forces through forward operations" and "conducting defence in depth". This meant that U.S. and Allied forces would not wait for Soviet units to enter the Atlantic but would instead engage them in the Norwegian and Barents Seas. The Maritime Strategy went even further than CONMAROPS by advocating for the direct prosecution of Soviet SSBNs in their "bastions"—highly defended waters close to the Kola Peninsula—as a means of influencing Soviet strategic calculations and conventional deterrence.

Carrier Operations in the Norwegian Fjords

One of the most distinctive tactical developments of the 1980s was the operation of aircraft carrier battle groups (CVBGs) within the confined waters of the Norwegian fjords. This tactic was a direct response to the threat of Soviet long-

An EA-6B Prowler of VAQ-135 "Black Ravens" and an F-14A Tomcat of VF-33 "Starfighters" are shown patrolling the waters off the coast of Norway during the 1985 deployment to the Norwegian Sea (Norlant operating area).

S-3A Viking (159404) of VS-24 "Scouts" on the August 15, 1986 to October 16, 1986 deployment aboard USS Nimitz (CVN-68) to the Norlant. The importance of anti-submarine warfare aircraft in the High North cannot be overstated, and the carrier-based Lockheed S-3A Viking proved its value while operating in the unforgiving conditions of the GIUK Gap.

range, missile-carrying aircraft, particularly the Tu-22M Backfire bomber. By operating in the fjords, the Navy sought to "harness geography and terrain for operational and strategic advantage".

Masking and Deception in Vestfjorden

During major exercises like Ocean Safari 1985 and 1986, the NATO Striking Fleet demonstrated its capacity to stage carrier flight operations within the Norwegian fjords, specifically Vestfjorden, which lies north of the Arctic Circle near Bodø. The steep fjord walls provided a natural physical barrier that shielded the carriers from prying Soviet radar and visual surveillance. This tactic allowed carriers like the to evade attack while simultaneously launching warplanes into the mission area.

To maintain this stealth, the fleet practiced stringent controls on electromagnetic emissions (EMCON). By "going dark," the carriers could mask their whereabouts even when operating in the Soviet "backyard". Anecdotal accounts from these operations describe F-14 Tomcats launching from the decks of carriers deep inside the fjords and staying below the cliffs until they reached the open sea, only then climbing to intercept Soviet bombers. This approach forced the Soviets to realise that their surveillance networks could be bypassed, creating significant uncertainty for their naval commanders.

The 1985 Vestfjorden Operations

The 1985 deployment of the USS *America* (CV-66) to Vestfjorden marked an important milestone. The aircraft carrier was part of a larger task force that also included the battleship USS *Iowa* (BB-61) and the cruiser USS *Ticonderoga* (CG-47). Participants later described the mission as involving "simulated attacks on Russia," with Soviet intelligence-gathering vessels ("trawlers") tracking the fleet as it crossed the Atlantic. During the operation, the *Iowa* and *America* carried out a series of coordinated manoeuvres. They divided into smaller groups—Surface Action Groups (SAGs) and Carrier Battle Groups—positioning themselves north of convoy routes while remaining close enough to quickly intercept potential threats. These exercises demonstrated that the Norwegian coastline could effectively serve as a natural defensive advantage—the "terrain of the sea"—helping to shield and support the Navy's most valuable assets.

ASW Tactics

Anti-submarine warfare (ASW) remained a "naval core competency" in the 1980s, as the Navy sought to neutralise the threat posed by the massive Soviet submarine fleet. The decade saw a transition from a reliance on passive acoustic detection to a more multifaceted approach as Soviet submarines became quieter.

The Navy's undersea surveillance system, originally called SOSUS and later expanded into IUSS in 1985, used fixed and mobile sonar arrays to track Soviet submarines. It detected low-frequency sounds, which analysts studied to identify submarine signatures and estimate their

An F-14A Tomcat (159452) of VF-102 "Diamondbacks" is seen during the deployment aboard USS America (CV-66), which lasted from August 24 to October 9, 1985. USS *America*, with CVW-1 embarked, was operating in the Norlant, where encounters with Soviet bombers and maritime patrol aircraft were a common occurrence during operations in the GIUK Gap.
(PHOTO: OFFICIAL US NAVY PHOTO BY LT DAVE PARSONS, USN)

An F-14A Tomcat (BuNo unknown) of VF-31 "Tomcatters" is seen preparing to launch from USS Forrestal (CV-59) in September 1987, during the deployment to the Norlant from August 27 to October 8, 1987. In the background, the mountains surrounding Vestfjorden in Norway are clearly visible. During the late 1970s and throughout the 1980s, increasing emphasis was placed on maintaining a continuous presence in the North Atlantic, driven by the growing activity of Soviet submarines and other naval vessels in the region.
(PHOTO: OFFICIAL US NAVY PHOTO BY PH3 BUCKNER)

positions. This information guided anti-submarine forces for interception, and the system remained highly secret—even within much of the Navy.

The S-3 Viking and the P-3 Orion

The S-3 Viking was a central component of the Navy's carrier-based anti-submarine warfare (ASW) capabilities in the 1980s, providing close-in protection for carrier battle groups. Considered a major technological leap, it delivered many of the same capabilities as the larger, land-based P-3 Orion, but in a compact, carrier-operated platform. Its highly integrated systems allowed the four-person crew to process and share data from multiple sensors in real time—an essential advantage when detecting and classifying faint submarine signals in complex environments.

Beyond its primary ASW role, the S-3 Viking also expanded into missions such as surface surveillance and electronic intelligence gathering, making it a versatile and indispensable asset aboard aircraft carriers. While the P-3 Orion remained the Navy's main long-range ASW aircraft, the Viking played the critical frontline role in protecting the fleet at sea.

Arctic Challenges

Operating within the GIUK Gap during the 1980s transformed the aircraft carrier from a blue-water asset into a specialised cold-weather platform. To project power toward the Norwegian Sea, carrier strike groups had to overcome an environment that was often more hostile than the Soviet adversary. As the Maritime Strategy of the decade shifted from avoiding the far north to operating there routinely, carrier aviation had to adapt to extreme and unfamiliar conditions.

One of the most serious challenges was topside icing, which threatened both flight operations and equipment. Ice accumulation could freeze catapult tracks, arrestor wires, and elevators, effectively preventing aircraft from launching or recovering. This made the carrier temporarily incapable of performing its primary function. Helicopter operations were especially vulnerable, and systems like the SH-60B LAMPS MK III depended on effective de-icing to function.

Visibility posed another major obstacle, particularly in the Marginal Ice Zone—the area between open water and solid pack ice. In regions like the Greenland Sea, fog and low cloud cover were frequent, with conditions obscured more than half the time during operations in 1985. These conditions eliminated the possibility of standard visual landings and forced pilots to rely on instrument approaches, often in poor weather and on unstable decks, long before the reliability of modern GPS systems.

Weather in the GIUK Gap was further complicated by the presence of polar lows—small but powerful low-pressure systems that developed rapidly. These storms could bring sudden winds exceeding 60 knots and were difficult to predict with the forecasting tools available at the time. As a result, aircraft could be caught in flight with little warning, forced to land on decks pitching violently in extreme conditions.

USS *Nimitz* (CVN-68) is seen off the Norwegian coast during Exercise Northern Wedding '86. The exercise was designed to test NATO's response to a potential Warsaw Pact invasion of Northern Europe. CVW-8 was embarked aboard USS Nimitz, on the deployment that lasted from August 15 to October 16, 1986.
(PHOTO: OFFICIAL US NAVY PHOTO BY PH1 BENO)

68

F-14A Tomcat (161608) of VF-21 "Freelancers" on the April 11, 1987 to October 13, 1987 deployment aboard USS Constellation (CV-64) to the Westpac., Indian Ocean and Persian Gulf. On August 8, 1987, two "Freelancers" aircraft fired missiles at an Iranian F-4E that was engaging a U.S. Navy P-3 Orion. The aircraft flown by Lt. Ferran was NK202/161607, which would have appeared identical to this aircraft except for its different modex number and BuNo.

INDIAN OCEAN AND PERSIAN GULF OPERATIONS

The decade of the 1980s represented a fundamental realignment of United States naval strategy in the Middle East, transitioning from a reliance on regional proxies to a sustained, direct presence of carrier battle groups. This shift was catalysed by the 1979 Iranian Revolution, which dismantled the "second pillar" of the Nixon Doctrine—the reliance on the Shah of Iran to secure Western interests in the Persian Gulf. As the decade opened, the United States found its naval forces increasingly tethered to the North Arabian Sea and the Gulf of Oman, a geographic commitment that would redefine the operational tempo of the U.S. Navy for a generation. The Carter Doctrine, articulated in early 1980, established that any attempt by an outside force to gain control of the Persian Gulf would be viewed as an assault on the vital interests of the United States and would be repelled by any means necessary. While the initial focus was on deterring Soviet expansion following the invasion of Afghanistan, the immediate reality for naval aviators involved the dual challenges of the Iranian Hostage Crisis and the erupting Iran–Iraq War.

The birth of Gonzo Station

While operations in the Persian Gulf and Gulf of Oman were nothing new in the 1980s, tensions were rising at the turn of the decade. The Iranian hostage crisis was in full swing and an attempt to rescue the 53 hostages was made on April 24, 1980, during Operation Eagle Claw. This operation involved US Navy carriers but not aircraft from the embarked airwings. Eight RH-53D Sea Stallion helicopters, drawn from the carrier USS *Nimitz* (CVN-68), were tasked with a complex multi-stage insertion into the Iranian desert. The mission suffered from a "haboob"—a massive, unpredicted dust cloud that blinded pilots and caused mechanical failures. Following a series of equipment malfunctions and a tragic collision at the "Desert One" refuelling site, which killed eight servicemen, the mission was aborted.

The psychological and operational impact of the Desert One disaster cannot be overstated. It forced a re-evaluation of how carrier air wings could support unconventional warfare. The failure was not one of pilot skill but of integrated planning and environmental intelligence. In the wake of the rescue attempt, the U.S. Navy established a semi-permanent presence in the North Arabian Sea at a location that became colloquially known as "Gonzo Station" (Gulf of Oman Naval Zone of Operations). The station allowed carriers

Lt. Cdr. Robert Clement fired an AIM-7 Sparrow missile at an Islamic Republic of Iran Air Force F-4E Phantom II after the aircraft had demonstrated hostile intent toward a U.S. Navy P-3 Orion operating over the Persian Gulf. The Sparrow struck its target but failed to bring the aircraft down.

E-2C Hawkeye (162619) of VAW-113 "Black Eagles" on the April 11, 1987 to October 13, 1987 deployment aboard USS Constellation (CV-64) to the Westpac., Indian Ocean and Persian Gulf. Close coordination between the different squadrons of the air wing was essential to success in routine operations, exercises, and combat missions alike. The engagement on August 8, 1987, served as a clear example of the effective cooperation between the airborne early warning squadron and the fighter assets of the air wing. In this instance, their coordinated response successfully protected the Lockheed P-3 Orion, deterring the attacking McDonnell Douglas F-4E Phantom II and ensuring that the U.S. Navy aircraft remained unharmed—the best possible outcome.

to remain within striking distance of Iran while staying out of the confined and easily mined waters of the Persian Gulf. The environmental and logistical strain of this commitment was unprecedented, as ships were required to stay on station for months at a time without port visits.

The Tanker War

As the ground war between Iran and Iraq settled into a bloody stalemate, both belligerents sought to internationalise the conflict by targeting the other's economic lifeline: oil exports. This maritime component, known as the "Tanker War," began in earnest in 1981 when Iraq declared a prohibited war zone in the northern Gulf. Iraq utilized its air force, specifically Mirage F-1s and Super Frelon helicopters armed with French-made Exocet anti-ship missiles, to target tankers loading at Iran's Kharg Island terminal.

Initially, Iraq's aim in attacking Iranian shipping was to provoke the Iranians to retaliate with extreme measures, such as closing the Strait of Hormuz to all maritime traffic, thereby bringing about foreign intervention against Iran. The United States had threatened several times to intervene if the Strait were closed. Lacking sophisticated anti-ship missiles in the early years, the Islamic Republic of Iran Air Force (IRIAF) adapted land-attack weapons for maritime strikes. F-4E Phantom IIs were observed using AGM-65 Maverick missiles and even AS-12 anti-tank missiles against the massive superstructures of oil tankers. While these smaller warheads rarely sank the vessels due to the massive size of oil tankers, they caused significant damage to bridge areas and accommodation blocks, resulting in high casualty rates among civilian seafarers.

The USS *Stark* incident

Naval aviation during the mid-decade period focused on surveillance and "show of force" missions. E-2C Hawkeyes and P-3C Orions provided the primary "eyes" for the Middle East Force, tracking the movement of both Iranian and Iraqi strike aircraft and ships. The goal was to prevent the closure of the Strait of Hormuz, a strategic chokepoint through which a significant portion of the world's oil supply flowed.

The policy of "neutral monitoring" collapsed on May 17, 1987, when an Iraqi Mirage F-1 fired two Exocet missiles at the frigate USS *Stark* (FFG-31). The attack killed 37 sailors and wounded 21 others, nearly sinking the ship. Although the Iraqi government apologised, claiming the pilot mistook the *Stark* for an Iranian tanker, the incident highlighted the extreme danger posed by modern anti-ship missiles and the inadequacy of the existing Rules of Engagement (ROE). The *Stark* was unable to defend itself effectively because its radar systems had not been locked onto the incoming aircraft, as the ROE at the time prioritized de-escalation.

In the aftermath of the *Stark* attack, the Reagan administration authorised a more aggressive posture. U.S. warships were permitted to challenge and, if necessary, engage any aircraft or vessel demonstrating hostile intent, rather than waiting for a hostile act to occur. This policy shift coincided with a request from Kuwait to protect its tankers from Iranian attacks. Iran had begun deploying Chinese-made CSSC-2 Silkworm missiles on the Faw Peninsula, providing a heavy, land-based anti-ship threat that outclassed previous Iranian capabilities. The U.S. agreed to "reflag" eleven Kuwaiti tankers as *American* vessels, making them eligible for U.S. Navy protection under the moniker Operation Earnest Will.

Reflagged Kuwaiti tankers are seen being escorted by USS Thach (FFG-43) in November 1987 during what became known as the "Tanker War."

Operation Earnest Will

Operation Earnest Will, which commenced in July 1987, was the largest naval convoy operation since World War II. It required a sophisticated layering of air and sea power. While frigates and destroyers provided close-in protection for the tankers, carrier battle groups on Gonzo Station provided "over-the-horizon" support. Carrier-based aircraft, including F-14 Tomcats and E-2C Hawkeyes, maintained a constant presence over the Strait of Hormuz to deter IRIAF intervention. The operation began on 23 July 1987, with USS *Crommelin*, USS *Fox*, and USS *Kidd* among the first escorts.

The heightened readiness of the fleet during Operation Earnest Will led to several near-misses and direct confrontations in the air. On August 8, 1987, a U.S. Navy E-2C Hawkeye and the cruiser Valley Forge detected two aircraft taking off from Bandar Abbas, Iran. Based on their flight profile, these were identified as Iranian F-4 Phantoms heading toward a U.S. Navy P-3C Orion surveillance plane. Two F-14A Tomcats from VF-21 "Freelancers", operating from the USS *Constellation* (CV-64), were vectored to intercept the bogeys. The Tomcats were led by Lt. Cdr. Robert Clement, with Lt. William "Bear" Ferran as his wingman.

Ferran's radar intercept officer initially made contact with the Iranian fighters on the AWG-9 radar at a range of 35 miles. Given the heat and humidity of the Gulf, a dense overcast hindered visual identification. The rules of engagement dictated that the F-14s had to wait until the bogeys were within 10 miles before firing. Believing the P-3 was under imminent threat, Ferran fired an AIM-7 Sparrow missile, which malfunctioned and fell into the water. Clement then fired a Sparrow, followed by another from Ferran. Clement witnessed a detonation and ordered a break turn to egress.

Throughout the engagement, the lack of visual confirmation led to significant internal debate. This led to criticism of the F-14 crews by their superiors, who accused them of 'firing Sparrows at false radar echoes caused by bad weather.' However, available evidence suggests that this accusation was unfounded. While some senior officers concluded the radar track might have been an Iranian C-130 Hercules, an E-2 controller involved in the vectoring later maintained that there was no mistake it had been an F-4E, even if a C-130 Hercules was also in the vicinity. A photo has since emerged of an Iranian F-4E Phantom II parked in a shelter, displaying damage to the aft fuselage and stabilators, supporting the air crew's version of the story.

The heightened tension of the "Tanker War" culminated in a catastrophic error on July 3, 1988. The Aegis cruiser USS *Vincennes* (CG-49), while engaged in a skirmish with Iranian gunboats, misidentified an Airbus A300 (Iran Air Flight 655) as an Iranian F-14 Tomcat. The aircraft had taken off from Bandar Abbas and was climbing within a recognised commercial air corridor. The *Vincennes* crew, under the stress of combat and influenced by "unconscious distortion of data," believed the aircraft was descending and accelerating toward the ship.

Captain William C. Rogers III authorised the launch of two Standard missiles, which destroyed the airliner and killed all 290 people on board. While the U.S. government characterised the incident as a tragic accident, it had a profound strategic impact. Believing the shootdown was a deliberate signal that the U.S. was about to enter the war on Iraq's side, the Iranian government finally accepted the UN-brokered ceasefire in August 1988.

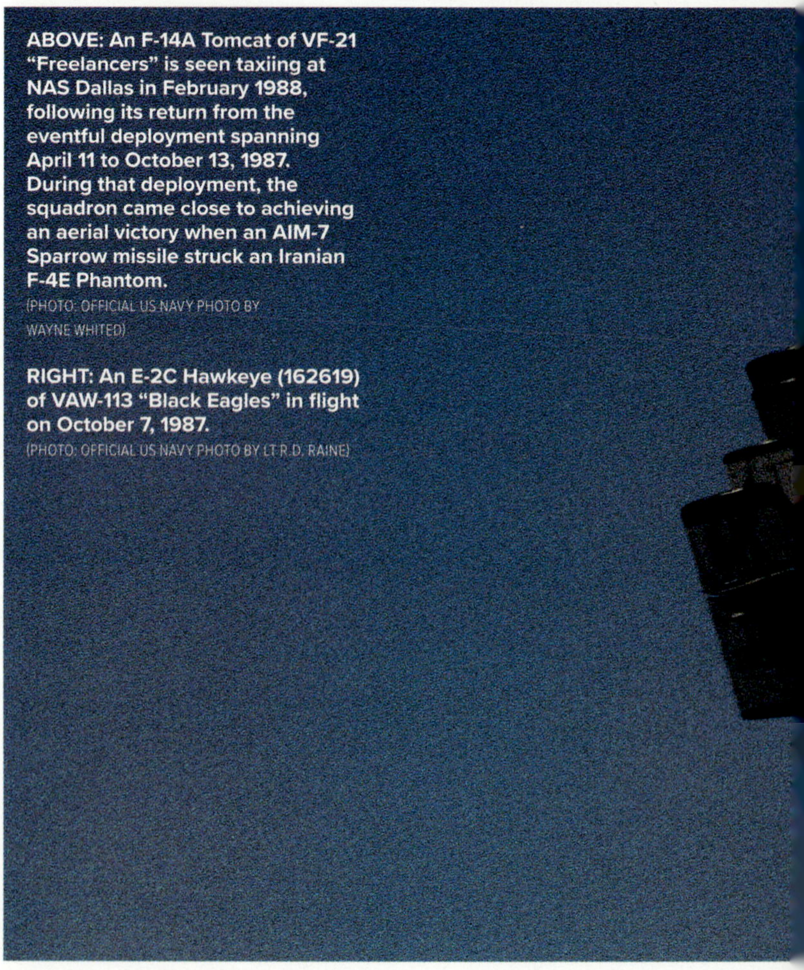

ABOVE: An F-14A Tomcat of VF-21 "Freelancers" is seen taxiing at NAS Dallas in February 1988, following its return from the eventful deployment spanning April 11 to October 13, 1987. During that deployment, the squadron came close to achieving an aerial victory when an AIM-7 Sparrow missile struck an Iranian F-4E Phantom.
(PHOTO: OFFICIAL US NAVY PHOTO BY WAYNE WHITED)

RIGHT: An E-2C Hawkeye (162619) of VAW-113 "Black Eagles" in flight on October 7, 1987.
(PHOTO: OFFICIAL US NAVY PHOTO BY LT R.D. RAINE)

206

A-6E Intruder (159574) of VA-85 "Black Falcons" on the June 7, 1982 to November 16, 1982 deployment aboard USS Forrestal (CV-59) to the Mediterranean Sea and Indian Ocean. The presence of four carrier groups in the eastern Mediterranean Sea during the crisis involving Lebanon created significant logistical strain on carrier operations. This A-6 is depicted carrying only a partial bomb load, reflecting the ordnance shortages that affected air wing strike capabilities at the time.

OPERATIONS SHORT OF WAR

Intervention in Lebanon

The deployment of United States naval power to the Eastern Mediterranean in 1982 marked one of the most significant demonstrations of carrier-based diplomacy and amphibious integration in the post-Vietnam era. As the Lebanese Civil War intersected with the regional ambitions of Israel and Syria, the United States Navy undertook a complex mission involving deterrence, evacuation support, and peacekeeping facilitation. Central to this effort was the presence of four aircraft carriers (USS *Forrestal* (CV-59), USS *Independence* (CV-62), USS *John F. Kennedy* (CV-67) and USS *Dwight D. Eisenhower* (CVN-69)) whose embarked air wings provided a continuous electronic and kinetic protective umbrella over the 32nd Marine Amphibious Unit (MAU) during operations along the Lebanese coast.

Regional stability shifted dramatically on June 6, 1982, when Israel launched Operation Peace for Galilee. Although publicly framed as an effort to create a buffer zone protecting northern Israel from PLO rocket attacks, the rapid advance of Israeli forces toward Beirut suggested broader strategic intentions. Israeli armour bypassed key coastal cities while the Israeli Air Force executed a highly effective suppression campaign against Syrian surface-to-air missile networks in the Bekaa Valley. The Reagan administration feared that a complete destruction of the PLO in Beirut could provoke direct Soviet involvement on behalf of Syria and Palestinian allies. To prevent escalation, the United States assembled a powerful naval task force of approximately fifty ships in the Mediterranean, forming the backbone of the Multi-National Force intervention.

Carrier Aviation as a Deterrent Shield

The F-14 Tomcat provided fleet defence and air superiority with its long-range AIM-54 Phoenix missile and powerful radar, allowing early detection of Syrian or Israeli aircraft approaching the task force or Marine landing zones. Although rules of engagement were strictly defensive, Tomcats routinely intercepted or monitored Syrian and Libyan aircraft operating near the operational area, serving as a visible deterrent without firing weapons.

Strike and close air support responsibilities rested primarily with the A-6E Intruder and A-7E Corsair II. The A-6E, equipped with the TRAM targeting system, offered precision strike capability in both night and adverse weather conditions, while the A-7E provided accurate daytime conventional attack support. Supporting aircraft such as the S-3 Viking conducted anti-submarine patrols and surface surveillance operations, monitoring Soviet naval activity and regional maritime movement during the crisis.

A key factor in the success of operations was the electronic surveillance and early warning infrastructure supporting the fleet. E-2 Hawkeyes maintained continuous airborne coverage over the Lebanese coast for extended periods, managing dense air traffic flows while monitoring Syrian radar emissions and coordinating potential strike responses. EA-6B Prowlers monitored the Syrian integrated air defence network and provided jamming capability, allowing U.S. naval forces to operate close to the Bekaa Valley despite its dense missile coverage.

The Beirut Evacuation

The arrival of the 32nd Marine Amphibious Unit in August 1982 marked a transition from naval presence to active peacekeeping intervention. Approximately 800 Marines landed as part of a multinational effort to oversee the evacuation of PLO forces from Beirut.

The PLO agreed to withdraw to several Arab countries. During this phase, naval

Lt. Cdr. Steve Weatherspoon and Lt. William "Skid" Massey are shown identifying the registration number of EgyptAir Flight 2843 by illuminating the markings on the aft fuselage of the airliner with a flashlight. This step was taken to ensure positive identification in the crowded skies over the Mediterranean Sea before initiating the diversion of the aircraft to NAS Sigonella in Sicily, which can be seen in the background.

An E-2C Hawkeye (161547) of VAW-125 "Tigertails," deployed from August 26, 1985 to April 16, 1986, is shown here preparing to launch from the deck of the USS *Saratoga* (CV-60) on January 28, 1986. This aircraft played a significant role during Operation Goldenrod, when it vectored F-14 Tomcats from CVW-17 toward EgyptAir Flight 2843. (PHOTO: OFFICIAL US NAVY PHOTO BY PH1 WILLIAM A. SHAYKA)

aviation focused primarily on surveillance and coordination. E-2C Hawkeye maintained a continuous airborne early warning presence while Marine helicopters supported liaison operations between command elements ashore and afloat. Naval aviation ensured that both Israeli naval forces blockading the port and Syrian air assets remained at a distance as approximately fourteen thousand PLO fighters departed Beirut. By early September the mission was considered complete and Marines began reembarking aboard amphibious ships.

The strategic situation deteriorated in mid-September following the assassination of President-elect Bachir Gemayel and the subsequent Sabra and Shatila massacres. International pressure prompted the return of multinational forces, and on September 29 the 32nd MAU landed again in Beirut. Unlike the earlier evacuation-focused deployment, Marines now occupied defensive positions around Beirut International Airport and faced artillery and sniper fire from surrounding high ground, marking a transition toward a longer-term peacekeeping mission.

Lessons learned

During this period the introduction of the Tactical Air Reconnaissance Pod System on F-14 Tomcats represented a major operational development. The Navy adapted the Tomcat to conduct photographic mapping missions over the Chouf and Aley mountains to locate artillery threatening Marine positions.

Sustaining four carrier battle groups in the Mediterranean imposed heavy logistical demands. The operational area off Lebanon required continuous replenishment at sea using fleet oilers and ammunition ships, while carrier onboard delivery aircraft maintained essential supply connections with shore bases in Sicily and Spain. A particular problem during this phase of the operation was a bomb shortage, which was due to the high strain on logistics.

The intervention also exposed weaknesses in joint-service coordination. Communication between Marine units ashore and Navy aviation units overhead was often hindered by incompatible equipment and fragmented intelligence sharing. Pilots faced particularly complex rules of engagement in a battlefield crowded with multiple militias using similar Soviet-origin equipment, making identification difficult and increasing the risk of accidental escalation. As a result, mission success was often measured by deterrence rather than combat engagement.

Operational lessons from Lebanon directly influenced planning for Operation Urgent Fury in Grenada the following year. While naval aviation in Lebanon had primarily performed protective and deterrent roles, carrier aircraft in Grenada conducted direct strike missions supporting amphibious assault forces.

Strategically, the deployment of four carrier battle groups to the Eastern Mediterranean marked a turning point in U.S. post-Vietnam military posture. It demonstrated renewed willingness to employ naval forces as instruments of regional stabilisation and signalled the growing importance of the Middle East as a focus of *American* military activity.

Granada: Operation Urgent Fury

Since the 1979 coup led by Maurice Bishop and the New Jewel Movement, Grenada had progressively aligned itself with the Soviet bloc and Cuba. This alignment was viewed by Washington as a direct challenge to the Monroe Doctrine and a potential threat to the stability of the Western Hemisphere.

The most provocative manifestation of this alignment was the construction

A-7E Corsair II (160554) of VA-87 "Golden Warriors" on the October 18, 1983 to April 11, 1984 deployment to the Caribbean and the Mediterranean Sea. This "Golden Warrior" is depicted carrying a mixed load of Mk 82 general-purpose bombs and CBU-59 APAM cluster bombs, configured for use on a close air support (CAS) mission during Operation Urgent Fury.

F-14A Tomcat (160904) of VF-103 "Sluggers" on the August 26, 1985 to April 16, 1986 deployment aboard USS Saratoga (CV-60) to the Mediterranean Sea. Lt. Cdr. Steve Weatherspoon and Lt. William "Skid" Massey´s F-14 Tomcat during the EgyptAir Flight 2843 incident

A Boeing 737-200 (SU-AYK) operated by EgyptAir was scheduled to fly from Cairo to Tunis on October 10, 1985. Among its passengers were the hijackers of the Achille Lauro hijacking. Aircraft from Carrier Air Wing 17 intercepted the airliner and forced it to land at Naval Air Station Sigonella, where the hijackers were taken into custody by U.S. special forces during Operation Goldenrod.

A flight of two Grumman A-6E Intruder aircraft from VA-176 "Thunderbolts" is shown airborne alongside other aircraft from CVW-6 on April 15, 1983, prior to Operation Urgent Fury. This mission marked the first combat employment of the A-6E TRAM Intruder.
(PHOTO: OFFICIAL US NAVY PHOTO BY LT. CDR SLOWIK)

of a 9,000-foot runway at Point Salines, facilitated by Cuban construction engineers and workers. United States intelligence agencies, including the Defense Intelligence Agency (DIA), identified that this airfield, ostensibly built for tourism, was over-engineered for commercial use. Analysts concluded that the facility was designed to support Soviet-made MiG-23 interceptors and long-range transport aircraft capable of interdicting vital sea lanes and air routes to Europe and the Middle East

Internal divisions within Grenada's People's Revolutionary Government erupted in October 1983 when a radical faction led by Bernard Coard, backed by the People's Revolutionary Army under General Hudson Austin, placed Prime Minister Maurice Bishop under house arrest. On October 19, supporters freed Bishop and marched to Fort Rupert, where PRA forces opened fire, killing civilians and executing Bishop and several ministers. A strict shoot-to-kill curfew followed. With nearly 1,000 *Americ*an medical students on the island and memories of the Iranian hostage crisis vivid, the Reagan administration concluded that rapid military intervention was necessary to protect U.S. citizens and restore stability.

Carrier Air Wing six operations

In response to the situation in Granada the US aircraft carrier USS *Independence* (CV-62) was sent to the area, with CVW-6 aboard.

VF-14 "Tophatters" and VF-32 "Swordsmen" were primarily tasked with establishing air superiority and preventing any intervention by the Cuban Air Force. Although the threat of Cuban MiGs did not materialise in the form of direct combat, the presence of F-14A Tomcats on constant Combat Air Patrol (CAP) ensured that the airspace over the eastern Caribbean remained uncontested.

Furthermore, the Tomcats played a critical role in reconnaissance. Using the Tactical Air Reconnaissance Pod System (TARPS), VF-32 provided high-resolution imagery of Grenadian defences and the movement of PRA forces.

The primary strike capability of CVW-6 was distributed among three attack squadrons. VA-15 and VA-87, flying the A-7E Corsair II, were the workhorses of the daylight Close Air Support (CAS) mission. The A-7E was highly regarded for its precision in delivering unguided ordnance, which was vital for the "surgical" bombing missions ordered by Metcalf to comply with the restrictive ROE. Operation Urgent Fury also marked the combat debut of the A-6E TRAM variant, flown by the Thunderbolts of VA-176 "Thunderbolts".

The EA-6B Prowlers of VAQ-131 "Lancers" were deployed to suppress enemy air defences (SEAD) and jam Grenadian communications. Although the PRA lacked advanced Surface-to-Air Missile (SAM) systems, they possessed numerous 12.7mm and 23mm anti-aircraft guns, some of which were radar-directed.

HS-15, known as the Red Lions, operated the SH-3H Sea King from the USS *Independence*. While their primary mission was anti-submarine warfare (ASW), the Red Lions were redirected to fly critical Search and Rescue (SAR) and Medical Evacuation (MEDEVAC) missions throughout the conflict. The SAR capability was tested early when four Navy SEALs tragically drowned during a pre-invasion night drop into the waters off Point Salines on October 23.

EgyptAir Flight 2843

The events occurring in October 1985, centred on the hijacking of the Italian cruise ship MS Achille Lauro and the subsequent military interception of EgyptAir Flight 2843, constitute one of the most complex and perilous diplomatic ruptures between the United States and Italy since the inception of the NATO.

Operation Goldenrod

By October 9, the hijackers had surrendered, and Egyptian President Hosni Mubarak publicly asserted that they had already left Egypt. This claim was quickly

The busy flight deck of USS *Saratoga* (CV-60) is seen during the deployment from August 26, 1985, to April 16, 1986. This eventful cruise culminated in operations leading up to Operation El Dorado Canyon, as well as involvement in the events surrounding EgyptAir Flight 2843. The two F-14A Tomcats in the foreground belong to VF-103 "Sluggers," which played a key role in these events. (PHOTO: OFFICIAL US NAVY PHOTO)

identified as a deception by U.S. and Israeli intelligence. Information relayed by Mossad and major general Uri Simhoni indicated that the four terrorists, along with leaders Abu Abbas and PLO official Hani el-Assan, were still at the Almaza airbase near Cairo, preparing to depart for Tunisia aboard an Egyptian commercial airliner.

The decision to intercept the Egyptian airliner, a mission dubbed "Operation Goldenrod," was developed by the National Security Council (NSC), with Oliver North playing a central role in coordinating intelligence between the Pentagon and the White House. Admiral James R. *Stark* proposed a tactical manoeuvre inspired by the World War II mission that intercepted Admiral Isoroku Yamamoto. The objective was to intercept the airliner over international waters and divert it to NAS Sigonella, where U.S. special forces could apprehend the terrorists. President Reagan gave the final authorisation for the operation to go ahead. USS *Saratoga* (CV-60), was positioned in the Ionian Sea and ordered to prepare a strike package. CVW-17 mobilised four F-14A Tomcats from the VF-103 "Sluggers" and VF-74 "Be-Devilers," supported by an E-2C Hawkeye, an EA-6B Prowler, and KA-6D tankers.

The mission's success depended on the precise identification of the aircraft among the heavy civilian traffic over the Mediterranean. Israeli intelligence confirmed the aircraft as an EgyptAir Boeing 737-200, flight number 2843, with the registration SU-AYK.

In the darkness of October 10, the F-14s operated with their external lights and radars off to avoid alerting the Egyptian crew. They were vectored by the E-2C Hawkeye toward potential radar contacts. After several false starts, the crew of the Tomcat piloted by Lt. Cdr. Steve Weatherspoon and Lt. William "Skid" Massey approached a contact 80 miles south of Crete.

Closing to within fifteen feet of the airliner, RIO Lt.Cdr. Dennis "Doc" Plautz utilised a handheld flashlight to confirm the EgyptAir livery and the tail number SU-AYK. Once the identification was confirmed, the Tomcats surrounded the airliner, remaining in its blind spots. CDR Zia, the mission commander on the Hawkeye, initiated radio contact with the Egyptian pilot. The Egyptian pilot initially attempted to continue toward Tunisia, which had already refused him landing rights under pressure from Washington. To shatter any remaining illusions of autonomy, CDR Zia issued the tactical command to the Tomcat pilots: "Lights on, now!". The sudden illumination of armed *American* fighters just feet from the cockpit windows forced the Egyptian crew to comply.

The Standoff at NAS Sigonella

The landing at Sigonella at 6:45 p.m. EST transformed a tactical success into a diplomatic crisis. The U.S. government had informed Italy of the interception only minutes before the aircraft entered Italian airspace, an act Craxi viewed as a blatant violation of sovereignty. As the 737 touched down, it was directed to the Italian sector of the base rather than the U.S. Navy sector.

The scene that unfolded on the runway was a "Mexican standoff" between two NATO allies. The Italian VAM and Carabinieri, acting on Craxi's direct orders, were the first to surround the 737. Moments later, two U.S. Air Force C-141 Starlifters landed, and fifty members of SEAL Team Six and Delta Force, commanded by Major General Carl Stiner, disembarked and formed a ring around the Italian airmen.

A critical technical element of the operation was the use of the Tactical Airborne Reconnaissance Pod System (TARPS) to document the standoff. Lt. Cdr. Larry Neal, a crew member of a VF-103 F-14A, provided a detailed account of taking infrared (IR) photographs of the scene from 4,000 feet, clearly showing the C-141 and Boeing 737-200 on the runway at NAS Sigonella. This event was highly dramatic in many regards, and led to US Navy F-14 Tomcats nearly engaging in dogfights with Italian F-104S ' and led to a significant diplomatic crisis between Italy and the US.

An E-2C Hawkeye (161781) of VAW-124 "Bear Aces" is seen taxiing at NAS Fallon on April 25, 1988. From this angle, most of the key visual identification markings are clearly visible. Starting at the nose, the modex number—602 in this case—indicates the aircraft's number within the squadron. Above the "NAVY" lettering, the name of the assigned aircraft carrier, USS *Theodore Roosevelt*, is displayed, with the squadron designation positioned below. On the vertical stabiliser, the tail code "AJ" signifies assignment to Carrier Air Wing 8 (CVW-8), and beneath it are the last four digits of the BuNo, sometimes referred to as the radio call number. With close inspection, the full BuNo can also be seen on the aft fuselage.
(PHOTO: OFFICIAL US NAVY PHOTO BY DOUGLAS D. OLSON)

ANATOMY OF A 1980s CARRIER AIR WING

Modex Number

The three-digit number, often referred to as the *Modex number,* was prominently displayed in multiple locations on the aircraft to ensure visibility from all angles. The first digit of the modex number indicated the squadron's position within the air wing. For example, the first fighter squadron used numbers in the 1xx range, the second in the 2xx range, the first attack squadron in the 3xx range, and so on.

The second and third digits denoted the aircraft's specific number within its squadron, typically ranging from 00 to 07 and then from 10 upward. The lowest numbers were assigned according to crew seniority. Aircraft bearing so-called "double nuts" numbers—such as 100, 200, or 300—were assigned to the Carrier Air Wing Commander (CAG). Numbers ending in 01 (e.g., 101, 201, 301) were typically assigned to the squadron Commanding Officer (CO), while 02-series numbers were assigned to the Executive Officer (XO), followed by other senior flight crew members.

Numbers containing 8 or 9 were omitted due to limitations in aircraft transponder systems, which used octal (base-8) coding and therefore could not accommodate these digits.

Modex numbers were always painted on the aircraft's nose, just aft of the radome. They were also applied to the upper surface of the trailing-edge flaps. When the flaps were lowered, as during launch

An F/A-18A Hornet (161954) of VFA-25 "Fists of the Fleet" is seen launching from USS *Constellation* (CV-64) on October 29 during carrier qualifications (CarQuals) for the February 20 to August 24, 1985 deployment. The modex number was not only displayed on the nose of the aircraft, but was also typically applied to the control surfaces or flaps of the wings and on the vertical stabilisers. This ensured that individual aircraft within the air wing could be easily identified from virtually any viewing angle. (PHOTO: OFFICIAL US NAVY PHOTO BY JO2 MILLIE TAMBERG)

and recovery operations, these numbers became visible from behind, facilitating aircraft identification by deck personnel. Additionally, either the full modex number or, more commonly, the last two digits were painted on the vertical stabiliser for further identification.

Tail Code

The two-letter code displayed on the tail of U.S. Navy aircraft—commonly referred to as the *tail code*—indicated the aircraft's Carrier Air Wing (CVW) affiliation. The first letter designated the fleet assignment: aircraft assigned to Atlantic Fleet air wings carried tail codes beginning with the letter **A**, while those assigned to Pacific Fleet air wings began with the letter **N**.

The table, right, lists each active Carrier Air Wing from the 1980s along with its corresponding tail code.

Bureau Number (BuNo)

The Bureau Number (BuNo) is a unique serial number assigned by the U.S. Navy Bureau of Aeronautics to every Navy and Marine Corps aircraft. This number was typically displayed on the aft fuselage. The Bureau Number system has been in continuous use since 1921, beginning with aircraft number 1 and continuing sequentially.

By the 1980s, most aircraft in active service bore BuNos in the 150000 to 169999 range. Older aircraft, such as certain variants of the A-3 *Skywarrior*, were produced earlier and often carried BuNos in the 140000 series.

Positioned above the BuNo on the aft fuselage was the aircraft's official designation, identifying its type and model.

Squadron Designation

The squadron designation was always displayed on the aircraft, typically in a prominent and clearly visible location. While its exact placement varied depending on the aircraft type, it was consistently positioned to ensure easy identification.

This alphanumeric designation indicated the squadron's primary mission role—for example, fighter (VF), attack (VA), or airborne early warning (VAW). The system originated in the late 1940s, when the carrier air wing structure was reorganised. Under this concept, Carrier Air Wing 1 (CVG-1) would be assigned squadrons such as VF-11, VF-12, VF-13, and VA-14. In this framework, the first digit represented the air wing, while the second digit indicated the squadron's position within that wing.

Although this system was never fully implemented—or at least not applied consistently—many squadrons active in the 1980s retained the designations assigned nearly forty years earlier. Fighter squadrons, which were typically the first two squadrons in an air wing, often carried designations ending in 1 or 2. This pattern was still evident in the 1980s in units such as VF-11, VF-21, VF-32, VF-41, and VF-111. Attack squadrons were more commonly assigned designations ending in 3, 4, 5, or 6, as seen in units such as VA-36, VA-55, and VA-83.

ATLANTIC FLEET		PACIFIC FLEET	
Tailcode	Wing	Tailcode	Wing
AB	CVW-1	NE	CVW-2
AC	CVW-3	NF	CVW-5
AE	CVW-6	NG	CVW-9
AG	CVW-7	NM	CVW-10
AJ	CVW-8	NH	CVW-11
AK	CVW-13	NK	CVW-14
AA	CVW-17	NL	CVW-15

As noted, there were significant deviations from this system, even as early as the 1950s. Nevertheless, elements of the original designation structure persisted throughout the 1980s.

From the 1950s onward, many new squadrons were established, including units with entirely new mission roles. These squadrons often had designations that did not follow the earlier numbering conventions. This was particularly true for Airborne Early Warning Squadrons (VAW) and Electronic Warfare Squadrons (VAQ), which were generally assigned consecutive numerical designations beginning at seemingly arbitrary points. Helicopter Anti-Submarine Squadrons (HS) also followed a consecutive numbering system, starting from -1.

Anti-Submarine Squadrons (VS) likewise used consecutive numbering, despite becoming part of the Anti-Submarine Warfare Air Groups (CVSG) in 1960, which could have prompted a redesignation. None of these squadrons were redesignated during that transition and instead retained their original numbering sequence, beginning at -20.

By the 1980s, many U.S. Navy squadrons had long and distinguished service histories and took considerable pride in their designations and heritage. Redesignations were relatively rare during this period. Even when squadrons transferred between air wings, their designations generally remained unchanged. Similarly, changes to squadron names and insignia became increasingly uncommon, though they were not entirely unheard of.

MAIN: An SH-3H Sea King (152121) of HS-2 "Golden Falcons" is seen in flight on February 27, 1987, during the deployment aboard USS *Kitty Hawk* (CV-63), which lasted from January 3 to June 29, 1987. Squadron designations were often prominently displayed on the aircraft, as seen here with "HS-2" positioned just below the "NAVY" lettering. Also note that the modex numbers used by helicopter anti-submarine squadrons were sometimes limited to a single digit.
(PHOTO: OFFICIAL US NAVY PHOTO BY PH3 EDWARD MARTENS)

INSET LEFT: An S-3A Viking (159729) of VS-24 "Scouts" is seen parked at NAS Fallon in April 1986. This aircraft serves as an example of anti-submarine warfare squadrons, whose designations generally ran consecutively from VS-20 through VS-42, although reserve squadrons in earlier decades deviated from this pattern. Notably, this S-3A is equipped with a luggage pod mounted on a wing pylon.
(PHOTO: OFFICIAL US NAVY PHOTO BY MICHAEL GROVE)

It is difficult to present a definitive representation of a U.S. Navy carrier air wing in the 1980s; however, these illustrations of the squadrons comprising Carrier Air Wing Nine (CVW-9) during its deployment from September 2, 1988, to March 2, 1989, provide a close approximation of a typical configuration.

Since the 1950s, a carrier air wing would generally include two fighter squadrons—in this case, VF-211 "Fighting Checkmates" and VF-24 "Renegades," both equipped with the F-14A Tomcat. Even this late in the 1980s, fighter and attack roles were still performed by separate squadrons, although strike-fighter units had begun to emerge in some air wings.

The three attack squadrons consisted of VA-146 "Blue Diamonds" and VA-147 "Argonauts," both flying the A-7E Corsair II, along with VA-165 "Boomers," which operated the A-6E Intruder. The combination of A-6 and A-7 aircraft enhanced the overall effectiveness of the attack force, as these platforms were capable of all-weather operations, suppression of enemy air defenses (SEAD), and limited air-to-air

engagement, in addition to their primary ground-attack roles.

Airborne early warning was provided by the E-2C Hawkeyes of VAW-112 "Golden Hawks," continuing a mission that had long been a cornerstone of carrier air wing operations. Electronic warfare capabilities were supplied by EA-6B Prowlers from VAQ-138 "Yellow Jackets." A relatively recent addition to carrier air wings was an expanded anti-submarine warfare capability, represented here by VS-33 "Screwbirds" flying the S-3A Viking. SH-3H Sea Kings from HS-2 "Golden Falcons" also contributed to both anti-submarine warfare and search-and-rescue missions.

Some air wings operated one or more dedicated Carrier Onboard Delivery (COD) aircraft, while others—such as CVW-9—relied on COD aircraft based ashore within the air wing's operational area. Taken as a whole, the aircraft of CVW-9 reflect a clear departure from the colourful paint schemes of earlier decades. Although a few squadrons retained the light gull grey camouflage and higher-visibility markings, colour was largely absent from the majority of aircraft in this air wing.

606
NAVY
163030

NG
0594
702
USS NIMITZ
NAVY
VS-33
CVW-9

O
NAVY
HS-2
212575
2575
NG

The 1980s was a period of significant transition for the U.S. Navy. The decade saw a shift from traditional air wings, where each squadron had a specific and well-defined mission, to modern air wings utilising multirole aircraft. This evolution occurred as older aircraft carriers approached the end of their service lives and were replaced by larger, more sophisticated vessels.

The "three sisters"—of which only two remained by the 1980s—were among the carriers being phased out. While USS *Franklin D. Roosevelt* (CV-42) was retired in 1977, USS *Midway* (CV-41) and USS *Coral Sea* (CV-43) remained in service throughout the decade. Operating modern fighters on these aging hulls presented unique challenges. USS *Midway* featured upgraded catapults for the F-4S Phantom II, yet it remained incapable of supporting F-14 Tomcat operations. Meanwhile, USS *Coral Sea* lacked the . The introduction of the lighter F/A-18A Hornet allowed both ships to remain frontline assets. By placing the fleet's latest fighter aboard its oldest carriers, these illustrations represent the broader transformation of Navy air wings during this era.

For its final deployment, USS *Coral Sea* embarked the newly established CVW-13, operating in the Mediterranean from May 31, 1989, to September 30, 1989. This wing featured three Hornet squadrons, including VMFA-451 "Warlords," VFA-132 "Privateers," and VFA-137 "Kestrels." The Hornet's integration began in the mid-80s and would eventually spread across the entire fleet. With three (and on earlier deployments, four) Hornet units aboard, the need for dedicated light attack squadrons operating the A-7 Corsair II was eliminated. However, because the Hornet had yet to acquire full all-weather capabilities, the A-6E Intruder remained essential. Consequently, the wing included seven aircraft from two A-6E TRAM-equipped squadrons: VA-55 "Warhorses" and VA-65 "Tigers." While unusual, this configuration was not unique.

The remainder of the air wing followed the standard carrier operating structure of the period. Airborne Early Warning was provided by the E-2C Hawkeyes of VAW-127 "Seabats," while Electronic Warfare duties were carried out by the EA-6B Prowlers of VAQ-133 "Wizards."

Air wings operating from *Midway*-class carriers did not include a dedicated Anti-Submarine Warfare (VS) squadron. Although operating the S-3 Viking was technically feasible, several constraints—most notably limited deck space and logistical considerations—prevented its regular deployment.

As a result, while no S-3 squadrons were assigned to CVW-13, the air wing retained its anti-submarine capabilities through HS-17 "Neptune's Raiders," which operated the SH-3H Sea King.

VFA-137
407

00
AK
158051
NAVY
500

10
AK
155620
NAVY
510

407
AK
USS CORAL SEA
NAVY

AK
600
USS CORAL SEA
NAVY
VAW-127/CVW-13

604
AK
161776
NAVY
604

612
NAVY
HS-17
148981
USS CORAL SEA
AK
CVW-13
8981

USS *Midway* (CV-41), with CVW-5 embarked, is seen underway in February 1985. Serving as a forward-deployed aircraft carrier throughout the 1980s, USS *Midway* was homeported in Yokosuka, Japan. Both USS Midway and USS Coral Sea (CV-43) operated the F-4 Phantom well into the 1980s; however, neither carrier ever hosted F-14 Tomcats or S-3 Vikings as part of their air wings. Several operational limitations restricted the use of these aircraft types, contributing to the eventual decommissioning of the Midway-class carriers as the decade drew to a close.
(PHOTO: OFFICIAL US NAVY PHOTO)

CARRIERS OF THE DECADE

The Midway Class

The 1980s presented a unique challenge for the two oldest aircraft carriers in active service: USS *Midway* (CV-41) and USS Coral Sea (CV-43). The last Midway-class ship, USS *Franklin D. Roosevelt* (CV-42)—part of the trio often referred to as the "three sisters"—had been decommissioned in the late 1970s.

Physical and Infrastructure Constraints

The primary limitation of the *Midway*-class was its inability to operate the F-14 Tomcat.4 Several factors contributed to this incompatibility. Both Arresting Gear and Catapult, Hangar deck height and flight deck geometry and blast deflectors were insufficient for operation of some of the aircraft that would be mainstays of the 1980s carrier air wings, in particular the F-14 Tomcat.

Despite these limitations, the *Midway* and *Coral Sea* were modernised to operate the F/A-18 Hornet. The Hornet's smaller footprint and lower maintenance requirements made it the ideal aircraft for these "small" supercarriers.

Operational Service of the Midway Class

The USS *Midway* (CV-41) remained a vital asset in the Western Pacific, homeporting in Yokosuka, Japan. In September 1982, the *Midway* participated in NORPAC 82, a massive joint training exercise in the Northern Pacific. During this exercise, two F-14s from the VF-114 and VF-213 operating off USS *Enterprise* were forced to divert to the *Midway* due to a rapid deterioration of weather at their home carrier. This event, involving an "OK 2" wire landing on a ship with only three wires, became a legendary moment in naval aviation, proving that the *Midway* could recover the Tomcat in an emergency, even if it could not support regular operations.

The SLEP Program: Forrestal and Kitty Hawk-class

To achieve the 15-carrier goal, the Navy implemented the Service Life Extension Program (SLEP), a massive industrial effort to refurbish and modernise the Forrestal-class and Kitty Hawk-class ships.

The Technical Challenges of SLEP

The SLEP was far more than a standard overhaul; it was designed to extend the service life of a carrier by 15 years beyond its original 30-year design life. The work for each vessel took approximately 28 months at the Philadelphia Naval Shipyard.

USS *Kitty Hawk* (CV-63), with CVW-9 embarked, is seen underway on March 2, 1987, during the deployment from January 3 to June 29, 1987, which took the ship on a round-the-world voyage from San Diego to Norfolk. Although the Forrestal- and Kitty Hawk-class carriers were closely related, this angle highlights a key distinguishing feature: the Kitty Hawk class had two elevators positioned forward of the island, whereas the Forrestal class had two located aft of the island. The Kitty Hawk class consisted of USS *Kitty Hawk*, USS *Constellation* (CV-64), USS *America* (CV-66), and USS *John F. Kennedy* (CV-67), although the latter eventually evolved into a class of its own. (PHOTO: OFFICIAL US NAVY PHOTO BY PH2 T. HENSLEY)

The first ship to undergo SLEP was the USS *Saratoga* (CV-60), entering the yard in October 1980. The technical scope of the project was exhaustive, including upgrades to both interior surfaces, installation of advanced radar systems, upgraded missile defence systems and improved propulsion.

Economic and Political Debates over SLEP

The SLEP program was a subject of intense scrutiny by the Government Accountability Office (GAO). Initial studies questioned whether the work should be performed in public naval shipyards like Philadelphia or private yards like Newport News. Proponents of the Philadelphia yard argued for the retention of specialised civil service skillsets, while private industry advocates pointed to potential cost savings and efficiency gains. Ultimately, the Navy maintained the program at Philadelphia to stabilise the workforce and leverage the learning curve acquired across multiple hulls.

By the mid-1980s, the SLEP had successfully modernised the USS *Forrestal* (CV-59), USS *Saratoga* (CV-60), and USS *Independence* (CV-62), ensuring they could operate the latest generation of carrier-based aircraft, including the F-14 Tomcat and F/A-18 Hornet.

USS *Enterprise* (CVN-65), with CVW-11 embarked, is seen underway on September 17, 1986, during Operation Sea Orbit. Flanking the Enterprise are USS *Truxtun* (CGN-35) and USS *Arkansas* (CGN-41), both nuclear-powered guided missile cruisers. USS *Enterprise* was the only ship of its class and is easily distinguishable from other aircraft carriers of the 1980s by its distinctive and unconventional island configuration.
(PHOTO: OFFICIAL US NAVY PHOTO)

USS *Nimitz* (CVN-68), with Carrier Air Wing 8 (CVW-8) embarked, is seen returning to Norfolk on May 28, 1981, following a night landing exercise off the East Coast. This event tragically resulted in the loss of 14 lives when an EA-6B Prowler crashed on the flight deck. At first glance, Nimitz-class carriers can appear similar to the Kitty Hawk class; however, a key distinguishing feature is the leading edge of the angled flight deck, which is set at a sharper angle than that of the Kitty Hawk class. Throughout the 1980s, additional Nimitz-class carriers were constructed, and by the end of the decade four were operational: USS *Nimitz*, USS *Dwight D. Eisenhower* (CVN-69), USS *Carl Vinson* (CVN-70), and USS *Theodore Roosevelt* (CVN-71).
(PHOTO: OFFICIAL US NAVY PHOTO BY PH1 KEN BREWER)

The Nuclear pivot: The Enterprise and Nimitz Class

The 1980s marked the definitive shift toward nuclear propulsion as the standard for the U.S. Navy's front-line aircraft carriers. While the USS *Enterprise* (CVN-65) had proven the concept in 1961, the high cost of nuclear surface combatants led the Navy to cease building nuclear cruisers and destroyers after 1975. However, the strategic advantages of the nuclear-powered aircraft carrier (CVN) were deemed indispensable for the high-end conflict scenarios of the late Cold War.

Technological Advantages of Nuclear Propulsion

The Nimitz-class carriers, offered endurance that was limited only by the delivery of aviation fuel, munitions, and crew provisions. The absence of large air intake and exhaust trunks required for conventional oil-fired boilers freed up approximately 300000 square feet of internal space. This allowed CVNs to carry twice the amount of aviation fuel (JP-5) and 30% more ordnance than their conventional counterparts.

Logistically, the ability of a nuclear carrier to sustain speeds in excess of 30 knots indefinitely was critical for rapid response to global crises. In 1980, during the Iran Hostage Crisis, the USS *Dwight D. Eisenhower* (CVN-69) demonstrated this by remaining on station in the Indian Ocean for 251 consecutive days without a port call, a feat enabled by the nuclear powerplant's independence from fuel oil tankers.

Innovation in Construction: Modular Techniques

The construction of the Nimitz class in the 1980s saw a revolution in shipbuilding efficiency. The USS *Theodore Roosevelt* (CVN-71) was the first carrier to be assembled using modular construction. Rather than building the ship from the keel up in a traditional linear fashion, large modules weighing hundreds of tons were pre-fabricated in "lay-down" areas and then hoisted into the dry dock using a massive 900-ton gantry crane.

This modular technique, enabled by improved industrial management and heavy lift technology, reduced the construction time for the Theodore Roosevelt by 16 months. This methodology became the standard for all subsequent carriers, including the USS *Abraham Lincoln* (CVN-72) and USS *George Washington* (CVN-73), both of which were under construction or authorised during the 1980s.

AIRCRAFT CARRIER	CLASS	PERIOD	AIR WING	AREA OF OPERATION
USS *Midway* (CV-41)	*Midway*-Class	Throughout	CVW-5	Forward deployed (Westpac.)
USS *Coral Sea* (CV-43)	*Midway*-Class	Throughout	CVW-14[1] - October 1983 CVW-13[2] October 1983 -	[1]Westpac. [2]Med.
USS *Forrestal* (CV-59)	Forrestal-Class	Throughout	CVW-17[1] - January 1983 CVW-6[2] May 1985 -	[1,2]Med.
USS *Saratoga* (CV-60)	Forrestal-Class	Throughout	CVW-3 -August 1980 CVW-17 January 1983 -	[1,2] Med.
USS *Ranger* (CV-61)	Forrestal-Class	Throughout	CVW-2 Throughout (CVW-9 July 1983 - March 1984)	Westpac.
USS *Independence* (CV-62)	Forrestal-Class	Throughout	CVW-6[1] - April 1985 CVW-17[2] August 1988 -	[1,2]Med.
USS *Kitty Hawk* (CV-63)	Kitty Hawk-Class	Throughout	CVW-15[1] - December 1981 CVW-2[2] January 1984 - August 1984 CVW-9[3] June 1985 -	[1,2,3]Westpac.
USS *Constellation* (CV-64)	Kitty Hawk-Class	Throughout	CVW-9[1] - January 1983 CVW-14[2] February 1985 -	[1,2]Westpac.
USS *Enterprise* (CVN-65)	Enterprise-Class	Throughout	CVW-11[1] Throughout (CVW-10[2] July - August 1987)	[1,2]Westpac.
USS *America* (CV-66)	Kitty Hawk-Class	Throughout	CVW-11[1] - November 1981 CVW-1[2] April 1982 -	[1,2] Med., Norlant. and Indian Ocean.
USS *John F. Kennedy* (CV-67)	Kitty Hawk-class/ John F. Kennedy-Class	Throughout	CVW-1[1] - March 1981 CVW-3[2] October 1981 -	[1,2] Med. and Norlant.
USS *Nimitz* (CVN-68)	Nimitz-Class	Throughout	CVW-8[1] - July 1987 CVW-9[2] June 1988 -	[1,2] Med., Norlant. and Indian Ocean.
USS *Dwight D. Eisenhower* (CVN-69)	Nimitz-Class	Throughout	CVW-7 Throughout	Med., Norlant
USS *Carl Vinson* (CVN-70)	Nimitz-Class	April 1982 -	(CVW-8 April to June 1982) CVW-15 March 1983 -	Westpac., Indian Ocean
USS *Theodore Roosevelt* (CVN-71)	Nimitz-Class	January 1987	(CVW-1 January - February 1987) CVW-8 March 1988 -	Med., Indian Ocean

F-14A Tomcat in the single-tone Light Gull Grey (FS 16440) camouflage scheme introduced in 1979. While this illustration depicts the aircraft in high-visibility markings, several variations existed, most notably those featuring enlarged national insignia.

MODELLING 1980s US NAVY AIRCRAFT

Base Camouflage Scheme

The early 1980s marked a period of significant change in U.S. Navy aircraft paint doctrine. Not since the 1950s had there been such a dramatic shift in how carrier-based aircraft were painted.

From 1956 until the late 1970s, the standard scheme had been Light Gull Grey (FS 16440) over Insignia White (FS 17875). The first sign of change came in 1977, when the Navy eliminated the Insignia White undersides, leaving aircraft painted entirely in Light Gull Grey.

The original two-tone scheme had been based on the principle of countershading, a concept derived from nature. When viewed from below against a bright sky, a light underside helped the aircraft blend in; conversely, when viewed from above against the darker sea or terrain, the grey upper surfaces provided concealment. However, as aerial combat evolved and engagements increasingly took place at higher altitudes, a critical flaw became apparent.

In modern air combat, situational awareness is essential. The contrast between a light underside and a darker topside made it easier for adversaries to determine whether a U.S. Navy aircraft was turning "hot" (toward the threat) or "cold" (away from it). The adoption of a single-tone Light Gull Grey scheme mitigated this issue by eliminating this visual cue, though it reduced effectiveness in environments where countershading was advantageous.

By the early 1980s, experimental alternatives were already being tested. One notable example was the Ferris scheme, which incorporated disruptive patterns and even a false canopy painted on the aircraft's underside to confuse opponents about its orientation. While effective at long distances, this approach was ultimately abandoned for technical and operational reasons.

The single-tone Light Gull Grey scheme remained in service well into the 1980s and could still be observed during Operation Desert Storm, particularly among some F-14 *Tomcat* squadrons. Although typically paired with high-visibility markings, many squadrons reduced the size of standard stencils—such as ejection seat warning triangles and rescue arrows—or applied them in subdued dark grey. In some cases, squadron insignia were similarly toned down.

A flight of A-7E Corsair II aircraft from VA-66 "Waldos" is seen over the Mediterranean Sea in May 1983 during the deployment aboard USS *Dwight D. Eisenhower* (CVN-69), which lasted from April 27 to December 2, 1983. The flamboyant markings of the 1960s and 1970s had not entirely disappeared but were becoming increasingly rare throughout the 1980s. This photograph also highlights variations in the saturation and hue of the otherwise standardised Light Gull Grey colour. Notably, aircraft 303's base colour appears distinctly different from the rest of the formation.
(PHOTO: OFFICIAL US NAVY PHOTO)

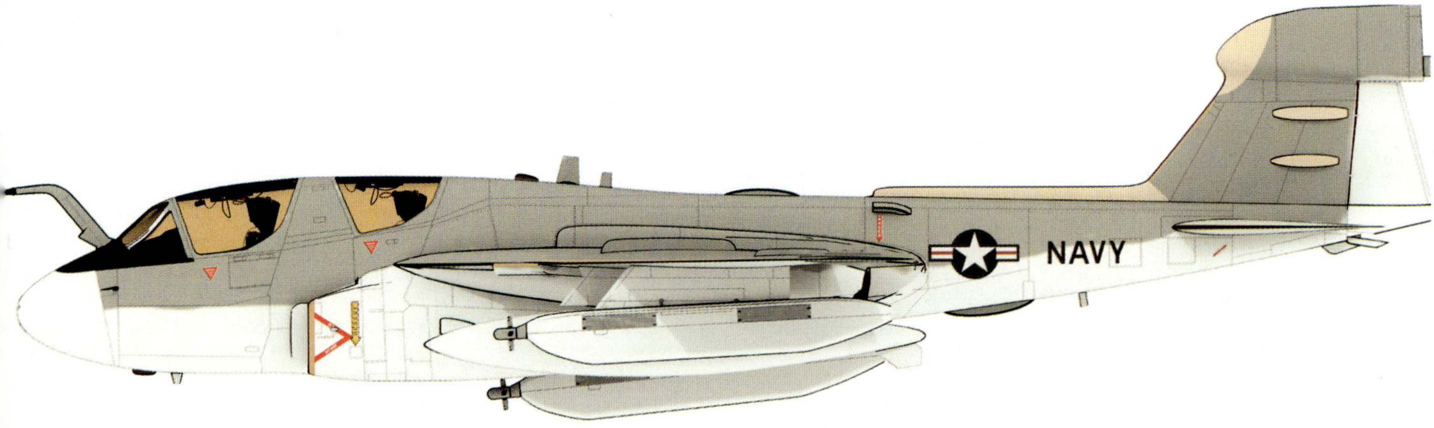

The EA-6B Prowler never carried the single-tone camouflage scheme; instead, it transitioned directly from the two-tone Light Gull Grey (FS 16440) over Insignia White (FS 17875) scheme to the Tactical Paint Scheme (TPS) illustrated elsewhere in this section.

E-2 Hawkeyes retained the two-tone Light Gull Grey (FS 16440) over Insignia White (FS 17875) scheme throughout the 1980s. As with other aircraft types, minor variations were observed; for example, the typically wavy camouflage demarcation line appeared straight on some aircraft. Overall, however, variation across the Hawkeye fleet remained limited.

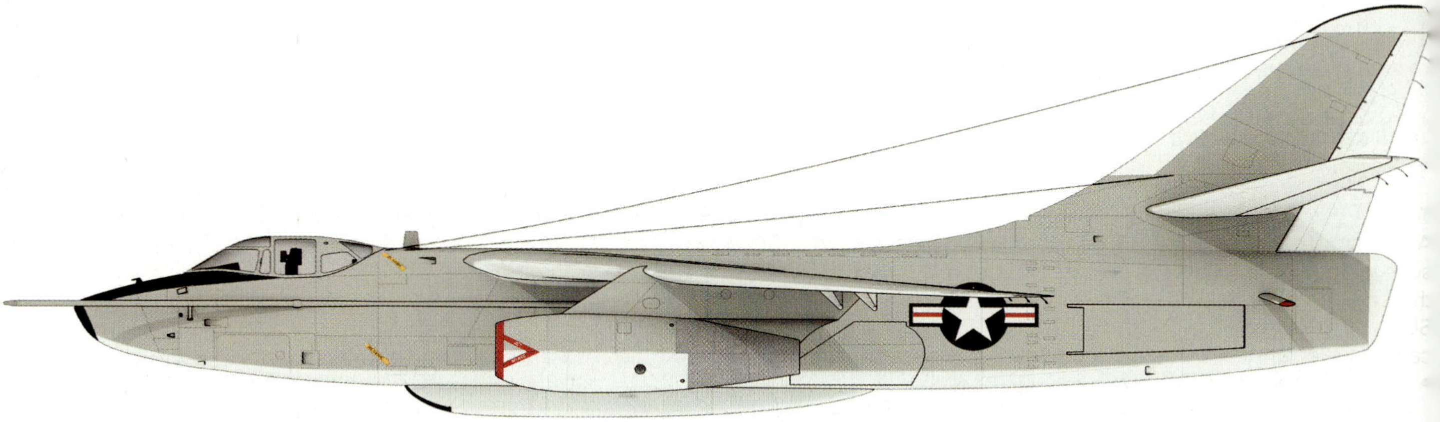

By the 1980s, only a small number of "Whales" remained assigned to Carrier Air Wings, and those still in service retained the Light Gull Grey over Insignia White paint scheme. Although some shore-based squadrons, such as VAQ-34, applied the Tactical Paint Scheme (TPS) to their Skywarriors, the carrier-based aircraft of VQ-2 continued to operate in this classic finish.

The S-3 Viking initially retained the Light Gull Grey over Insignia White camouflage scheme; however, as the decade progressed, Light Gull Grey was replaced by Dark Ghost Grey (FS 36320), and the national insignia was rendered in a single-tone grey. This transitional scheme was short-lived, and by the end of the decade the Tactical Paint Scheme (TPS) had also been applied to the S-3 Viking.

An F-14A Tomcat (BuNo unknown) of VF-211 "Fighting Checkmates" is seen in flight during the January 3 to June 29, 1987, deployment aboard USS Kitty Hawk (CV-63). This aircraft illustrates the difficulty of defining standard colours for the base camouflage scheme during the early years of the Tactical Paint Scheme (TPS). The patchwork appearance of repainted sections reveals variations in tone, with some areas appearing more bluish and others less saturated. It also demonstrates how paint repairs were often carried out using whatever materials were available, rather than perfectly matching the original base colour. (PHOTO: OFFICIAL US NAVY PHOTO BY PH3 D. W. HANSELMAN)

Tactical Paint Scheme (TPS)

A true successor to the single-tone scheme emerged in the early 1980s with the introduction of the Tactical Paint Scheme (TPS). This three-tone grey camouflage consisted of:

- FS 35237 – Medium Grey (upper surfaces and, in some cases, markings)
- FS 36320 – Dark Ghost Grey (vertical surfaces and mid-tones)
- FS 36375 – Light Ghost Grey (undersides)
- FS 36081 – Seaplane Grey (markings)

This scheme retained the principle of countershading but employed subtle gradations that made it far more difficult to determine whether an aircraft's upper or lower surface was visible. Under TPS regulations, all markings were applied in dark grey, marking the definitive end of the colourful squadron markings that had characterised the 1960s and 1970s (though these would later return in the 1990s).

By the early 1980s, squadrons began transitioning to TPS. Many *Tomcat*, *Intruder*, *Corsair II*, *Prowler*, and *Phantom II* units adopted the scheme during this period. Prior to full adoption, some units employed hybrid schemes, using a single TPS tone as an interim solution.

The Navy's newest aircraft, the F/A-18 *Hornet*, was delivered directly from the production line in the Tactical Paint Scheme and never wore any other camouflage.

An F-14A Tomcat (160902) of VF-84 "Jolly Rogers" is seen taxiing at NAS Fallon in April 1986, shortly before embarking aboard USS Nimitz (CVN-68) for a deployment to the North Atlantic (NORLANT). In contrast to the previous example, which displayed a patchwork of paint repairs in multiple shades of grey, this aircraft features a more uniform and disciplined camouflage scheme, closely resembling the standard specified in FED-STD-595. (PHOTO: OFFICIAL US NAVY PHOTO BY MICHAEL GROVE)

As the 1980s progressed, most F-14A Tomcats were repainted in the Tactical Paint Scheme (TPS). Although the scheme depicted here represents the standard specified in MIL-STD-2161, considerable deviations were observed across the Tomcat fleet. Some aircraft carried only two—or even a single—of the prescribed colours, often combined with variations in camouflage pattern, resulting in a wide range of documented configurations.

The F/A-18 Hornet rolled off the production line already finished in the Tactical Paint Scheme (TPS), which meant that their paint schemes—particularly in the early years—were generally consistent with the illustrations shown here.

Toward the end of the decade, SH-3 Sea King helicopters also began to receive a paint scheme similar to the Tactical Paint Scheme (TPS). This finish typically consisted of a single-tone base colour (FS 36320), with markings applied in FS 36081.

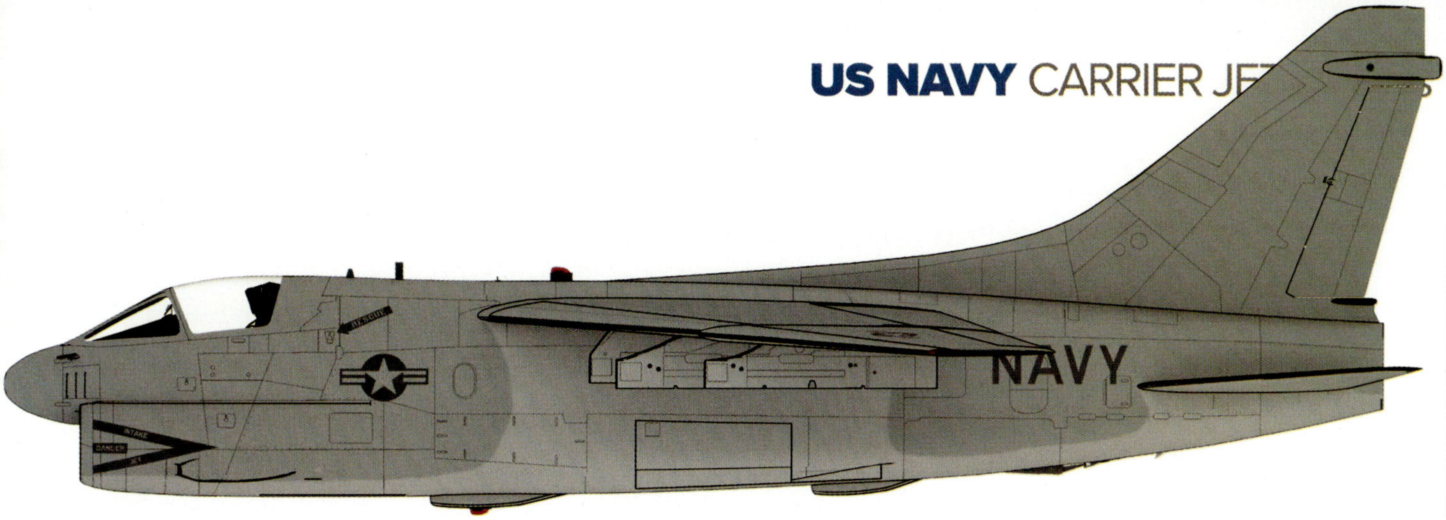

The A-7 Corsair II were all repainted over the course of their service lives—many of them multiple times—and, much like the Grumman F-14 Tomcat, they often displayed noticeable variations in their base camouflage schemes. The finish shown here represents the standard specified in MIL-STD-2161, although numerous examples of deviations from this specification can be found in service.

Like the A-7 Corsair II, the A-6 Intruder were repainted during their service lives from light gull grey over insignia white to the TPS, which resulted in noticeable variations in their base camouflage schemes. The scheme depicted here, however, follows the standard specified in MIL-STD-2161.

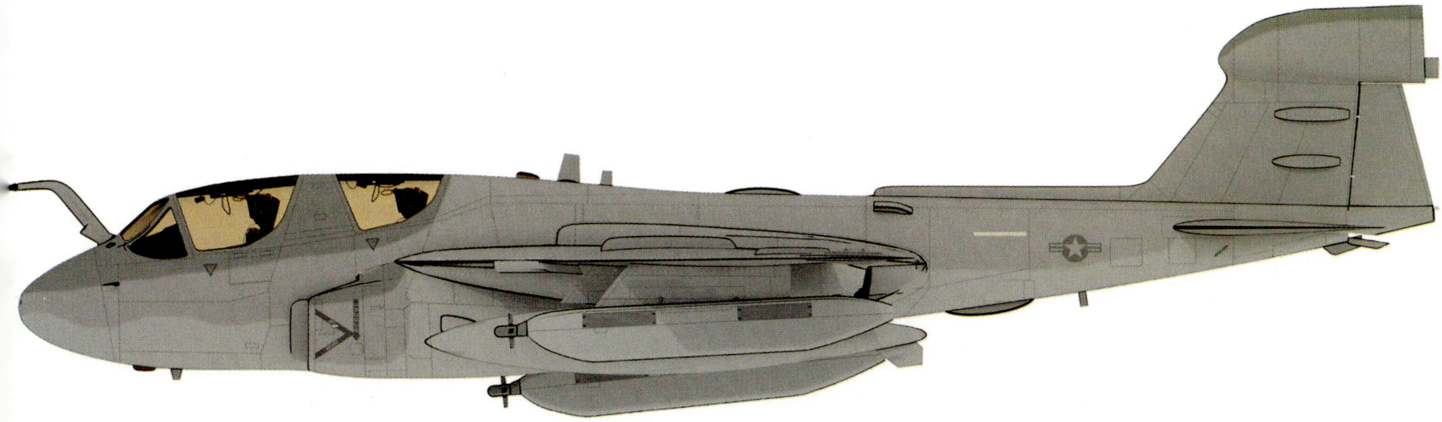

The EA-6B Prowler followed the same general pattern as the A-7 Corsair II and A-6 Intruder, typically carrying the standard Tactical Paint Scheme (TPS) illustrated here. However, the EA-6B fleet displayed considerable variation from this standard in both camouflage pattern and colour application. Some aircraft appeared in a single-tone scheme, while others carried two- or three-tone camouflage finishes.

Two Grumman F-14A Tomcat aircraft from VF-103 "Sluggers" are shown on the deck of the USS *Saratoga* (CV-60) during the August 26, 1985 to April 16, 1986 deployment to the Mediterranean Sea, prior to Operation El Dorado Canyon. Note the noticeable variation in base paint colour between the two aircraft from the same squadron—an illustrative example of the finish inconsistencies that were common among many squadrons during this period. (PHOTO: OFFICIAL US NAVY PHOTO)

Variations and Inconsistencies

Although MIL-STD publications provided strict camouflage standards, variations were common. For example, F-14 *Tomcats* were often seen in patterns using only two of the TPS colours. Additionally, there are numerous examples of single-tone schemes using either of the TPS coloursl as a base.

Further complicating matters, TPS colours exhibited significant variation. This was largely due to new Volatile Organic Compound (VOC) regulations, which required manufacturers to reformulate paints using early-generation urethane and polyurethane coatings. These formulations often resulted in inconsistencies.

Application methods also contributed to variation. Factors such as the number of coats applied and the degree to which the paint covered the underlying primer could significantly affect the final appearance.

Although the transition to TPS was rapid—resulting in largely low-visibility carrier air wings by the mid-1980s—some exceptions remained. E-2 *Hawkeye* squadrons retained the Light Gull Grey scheme throughout the decade, while the S-3 *Viking* initially used a transitional scheme combining a Light Gull Grey upper surface with an Insignia White underside before later adopting TPS.

Causes of Colour Variation

Significant variation in the hue of TPS paints during the 1980s resulted from formulation challenges, application inconsistencies, and environmental effects.

Technical and Formulation Challenges

VOC regulations: Environmental regulations introduced in the late 1970s and 1980s required reduced VOC content, forcing rapid reformulation of paints. Early urethane and polyurethane blends were still being refined, leading to inconsistencies.

Low-detectability requirements: TPS aimed to reduce both visual and radar detection, requiring multiple grey tones tailored to different aircraft profiles, making consistent colour reproduction difficult.

Manufacturer differences: Paint was sourced from multiple suppliers, and variations in chemical composition between batches contributed to differences in hue.

Application and Maintenance Factors

Application methods: As with earlier "haze paint" used during World War II, TPS effectiveness depended heavily on the skill of the painter. Variations in application techniques across maintenance depots often resulted in noticeable differences.

Touch-ups and repainting: Aircraft frequently underwent repairs and repainting, often using different paint batches, leading to visible patchwork effects.

Weathering: Early TPS coatings were less resistant to environmental exposure. Sunlight, UV radiation, and harsh weather caused uneven fading and chalking, altering the appearance of the paint over time.

Modelers and aviation enthusiasts have long noted these variations, with colours appearing anywhere from greenish or khaki tones to blue-grey depending on lighting conditions.

By the 1990s, advances in paint technology—including improved UV inhibitors and more refined urethane formulations—significantly improved consistency and durability, reducing the degree of variation seen in the previous decade.